Sex, Genes & Rock 'n' Roll

Sex, Genes & Rock 'n' Roll

How evolution has shaped the modern world

ROB BROOKS

University of New Hampshire Press
Durham, New Hampshire

University of New Hampshire Press
An imprint of University Press of New England
www.upne.com

Manufactured in the United States of America

First published in Australia by UNSW Press.

University Press of New England is a member of the Green Press Initiative. The paper used in this book meets their minimum requirement for recycled paper.

Library of Congress Cataloging-in-Publication Data

Brooks, Rob, 1970–
Sex, genes & rock 'n' roll: how evolution has shaped the modern world / Rob Brooks.—1st ed.
p. cm.
Includes bibliographical references and index.
ISBN 978-1-61168-236-6 (cloth: alk. paper)—ISBN 978-1-61168-237-3 (ebook)
1. Human evolution. 2. Social evolution. 3. Natural selection. I. Title.
GN281.B764 2012
599.93'8—dc23

2011046328

5 4 3 2 1

For Patti and Ben, my immediate ancestors,
Ben and Lily, my descendants,
And Jacqui, my partner.

If we're "above" nature, it's only in the sense that a shaky-legged surfer is "above" the ocean.

—Christopher Ryan and Cacilda Jethá

Ecology is the overall science of which economics is a minor specialty.

—Garrett Hardin

I believe in equity for all women and men in the world. Not equality. Equity is justice, freedom from bias, the opportunity to use your capabilities to their fullest. Equality is sameness, one mold.

—Irene Tinker

Contents

Prologue

I HAVE A DREAM JOB. At least it is what I might have considered a dream job in my adolescence; I am paid to think about sex. I am an evolutionary biologist who studies how natural selection has shaped the bodies and behaviors of animals, including humans. Admittedly most of my field observations and lab experiments are on insects and tropical fish, but it is still more fun than working in a bank, the pay is better than waiting tables and it is altogether more secure than playing in a rock-and-roll band. For evolutionary biologists—as for many adolescents—there are few subjects as compelling and mysterious as sex. That is because natural selection, the process by which evolution occurs, is all about reproduction.

Forget everything you have heard about "survival of the fittest." Survival is only a subplot in the grand story of evolution. Each animal, plant, fungus, bacterium and virus on earth today is here because its parents, grandparents and every one of its ancestors succeeded in having and raising at least one offspring. Natural selection is simply what happens when some individuals reproduce and others don't. It is a wasteful and inefficient way of designing things, but that is because it is not about design at all. All of the exquisite functionality in the living world, from the delicate wings of a dragonfly to our human capacity for language, is the by-product of one utterly boring fact: that in every generation of every species some individuals reproduce more successfully than others.

So it should not be a surprise that evolutionary biologists like me share a healthy and enthusiastic fascination for reproduction and for sex.

I am writing this book for many reasons, but foremost is the fact that evolution is both fascinating and important. My students, colleagues and I are motivated by a curiosity about why animals, including humans, live their lives as they do. In recent years we have discovered how the genes that make a male guppy attractive also make him more likely to die young, why male crickets do best on a high carbohydrate diet but females need a high protein diet, and that female dung beetles manipulate how hard their daughters-in-law work toward rearing the females' grandchildren. We have also used the tools of evolutionary biology to understand why left-handed cricket batsmen are so successful at the top international levels and why, in developing countries, women tend to be at greater risk of obesity than men.

The answers to all of these questions come down, at some level or another, to sex. Not always directly, but because sex and reproduction are so very central to evolutionary success, sex tends to change everything. The work that we do in my research group at the University of New South Wales in Sydney is just a tiny part of the work that scientists are doing in universities, museums, zoos and nature reserves around the world. All of this wonderful research contributes to an improved understanding of the world where we live, and of ourselves, and it is the wonder of this understanding that I want to share in these pages.

Evolution is not only fascinating; it is also profoundly important. Natural selection may be mundane, but according to the philosopher Daniel Dennett, it is also "the most important idea anybody ever had." That might appear to be a big claim, but it is a realistic one. After all, the very simple process of natural selection fashioned almost every aspect of the living world from human consciousness to the mold that grows on your bread. Unfortunately, only a small proportion of adults in even the most educated nations are likely to agree with Dennett. The number of wonderful books on evolution at your local bookstore may be growing, but it is easily outstripped by self-help manuals, dating advice, astrology, diet books and management babble, almost all of which are of dubious value, and some of which can be downright harmful. Speaking of harm, it is possible to spend more on Scientology courses and books than you could spend buying everything in the popular science section at a good bookstore without learning anything of lasting value other than the unfortunate fact that a fool and his money are soon parted.

Professionals who work with living organisms are also often oblivious about the processes that made those organisms what they are today. According to Randolph Nesse, one of the most eloquent advocates of Darwinian medicine, most medical schools in America and the United Kingdom have not one evolutionary biologist on their faculty, and most are reluctant to shoehorn teaching about evolution into an overloaded curriculum. But new insights from Darwinian medicine illuminate the origins of diseases like cancer and Alzheimer's disease, the ways in which our bodies fight infection, the basis of depression and schizophrenia, and the complex conflicts between mother, father and baby that cause some pregnancy complications. Throughout the history of medicine, most progress came from improved understanding of *how* we get infections, diseases and other mental and physical afflictions. But medicine can become even better when we understand *why* we get sick, and *why* our bodies, including our minds, respond to pathogens and stress in the ways that they do.

Important evolutionary insights are not restricted to medicine. Evolution is useful anywhere living organisms are involved, such as agriculture, fisheries, biotechnology, conservation, and carbon accounting. Most of all, evolution can teach us much about what it means to be alive, and why people do what they do. The new science of evolutionary psychology is generating so many insights into human behavior and motivation that it could easily be the most important development in psychology since Herr Professor Freud cracked open his notebook and asked for the first time "So, tell me about your childhood."

The idea of this book is not to give a comprehensive or watertight lesson on evolution, but rather to provide a glimpse of the world through the eyes of an evolutionary biologist. Although most of my research career has been spent studying small animals, recently my students and I have started to test some of our ideas in humans. I enjoy these studies enormously because my own interest in evolution comes from a desire to understand both human history and the lives people lead today. This book gives me a chance to show you how an evolutionary perspective can give useful and interesting insights into familiar issues and problems. This is not a book that aims to solve a single problem or make a single point. I have simply chosen to explore a few topics to provide a tasting platter of ideas. If you want to know more about any of these topics, each of which could fill several books, visit my website at <www.robbrooks.net> where I will continue writing about evolution in the modern world.

Recent history happened too

Modern humans evolved in environments much different from the ones most of us now inhabit. The environment to which we are all adapted is the environment or, more correctly, the range of environments, where our ancestors lived. Therefore it matters who those ancestors were and where they lived, died and reproduced, because those are the conditions that shaped who we are today.

Here is the minimum you need to know about human history in order to follow this book. Until somewhere between 5 and 7 million years ago, our ancestors were the same African forest-dwelling apes that also gave rise to chimpanzees (of which there are two modern species, the bonobo and the common chimpanzee). Since then, our ancestors became ever more human-like in every respect, though in fits and starts. Brains got bigger, posture became more erect, and skins got less hairy. Our ancestors also moved from being largely vegetarian apes to being omnivorous hunters of meat and gatherers of plant-based food by 2 million years ago. This creature—*Homo erectus*—was similar enough to modern people that a non-expert would have trouble telling them apart, and it spread out from Africa, covering much of Europe and Asia—at least as far as eastern China and Indonesia.

The African population of *Homo erectus* gradually gave rise to what we now consider our species, *Homo sapiens*, by at least 200,000 years ago. Everybody alive today traces their ancestry to a population of modern humans that lived in Africa relatively recently; probably around 60,000 years ago. Descendants of this group spread out across all Africa, into Asia Minor, and from there throughout the world. Year by year, people moved to find and use new hunting and fishing grounds, slowly peopling the planet. In places where *Homo erectus* or other pre-modern humans already lived, anatomically modern humans wiped them out, although in some cases, like the Neanderthals, there was some interbreeding.

The ancestors of every person alive today were African hunter-gatherers until 60,000 years ago, when modern humans spread out from Africa, but these ancestors undoubtedly lived in a vast range of habitats within Africa. Until 12,000 years ago at the earliest, all humans were still hunter-gatherers, inhabiting ever-greater portions of the world each year. But not all hunter-gatherer societies are alike. They differ enormously depending on the conditions and the kinds of food available where they live: from the !Kung of Africa's Kalahari desert, who eke an existence by hunting mammals and gathering plant matter, to the Inuit of the Arctic, who hunt marine mammals, catch fish and eat almost no plant matter at all.

In the last 12,000 years, people in various parts of the world learned to cultivate trees (horticulture) and then crops and to domesticate animals (agriculture). Horticulture, and then agriculture, imposed on humanity the most profound changes ever to take place in human history. This transition to agriculture was messy, chaotic and took hundreds or thousands of years. In some parts of the world almost everybody depended on agriculture by 6000 years ago. In other parts, farmers and hunter-gatherers lived near to one another for centuries. Some contemporary societies, such as those of the Amazon, cultivate trees but have never truly domesticated any plants or animals. In an even smaller number of contemporary societies, hunting and gathering remains the only form of subsistence those people have ever known.

While our ancestors spent the vast majority of history as hunter-gatherers, the ancestors of many people alive today have been farmers for millennia. Ignoring the profound changes that agriculture wrought on human life and the equally profound evolutionary consequences of those changes would be a mistake. For example, people whose ancestors domesticated animals that can be milked have since evolved the capacity to digest dairy products. And the descendants of ancestors who farmed have genetic pathways that allow them to detoxify alcohol more efficiently than people whose ancestors never farmed, because only farmers have excess grain to ferment into alcohol.

Tempting as it is to imagine our history as one long, sunny stretch of hunting and gathering on a uniform African savannah, we should remember that even within Africa the environments our ancestors lived in were bewilderingly variable. Just as the environments that people inhabit today are as different as the Atlas Mountains of North Africa, the crowded cities of eastern China, the icy wastes of Greenland and the lush rainforests of the Brazilian Amazon, so our ancestors lived in temperate rainforest, tropical atolls, Arctic tundra, grassland savannah and searing deserts. Even within each of these modern or ancient environments there are people whose immediate surroundings differ from one another in the food they get to eat, the quality of the water they drink, the density of people who live around them, and how hard they have to work to make a living. Next-door neighbors even differ in their social status and stress levels. Because it is quite possible for a person to grow up in any number of possible circumstances, the genes that shape our development and influence our behavior are often genes that work well in a range of environments. One story in human evolution that remains largely untold is the way natural selection has enabled us to deal with variation in our circumstances, and that is an important theme of this book.

Evolution's place in society

We have learned so much since Charles Darwin first explained natural selection 150 years ago, but the "most important idea that anyone ever had" remains poorly understood and underappreciated by most people. Most of the reasons for this fall into three groups: the ancient tensions between science and religion, uncertainties over how evolution meshes with other ways of understanding our world, and outdated views about how evolution works.

Conflict between faith and reason has been with us at least since recorded history began and it won't go away soon. People resolve or live with this conflict in any number of ways. Atheists suffer no internal conflict because they simply have no religious beliefs. Many religious people compartmentalize their religious faith and yet use reason and science to understand their world. For others, there is a need to reject reason because it conflicts with their strict adherence to pre-scientific accounts of the world. It disappoints me that peddlers of superstition continue, in the twenty-first century, to perpetuate Stone Age ignorance about how the world works. In so doing, they mislead the very people who depend on their guidance and leadership.

But society also loses out because scientists end up spending so much energy and time debunking antediluvian views rather than getting on with exploring how the world really works and explaining it to people who are genuinely curious. I therefore assume that the readers of this book are curious about the real world around them, and I avoid lengthy discussion of creation myths and the conflict between faith and reason. I give readers the credit to see that the really interesting insights that science has to offer can be discussed without being sucked into the tedium of dispelling outdated beliefs about how the world as we know it came to be.

The second reason that evolutionary thinking is not as prominent as it should be is that it is seldom clear how it fits with other ways of understanding our world. Natural selection made living organisms what they are today by acting over unimaginably long periods, and it can be tempting to think that that past no longer matters. Instead we focus on economic and cultural processes that are immediately visible to us. In this book I want to show how evolution is taking its place alongside the studies of economics and culture, to enhance how we understand our lives, human history, and the prospects for social improvement. The relationship between evolution and economics is a budding romance. The relationship between evolution and culture—by which I mean all those things we learn and acquire socially as we grow up and live our lives—is more of a long and tortured estrangement. It is not quite a divorce, and I am optimistic that there is hope for rapprochement if not complete reconciliation.

The flight from Sydney to New York, via Los Angeles, is an exercise in round-the-clock masochism. If everything goes well, you can check into some lackluster hotel about 24 hours after leaving home, but the date will not have changed. At the start of one such dreary journey in 2008, somewhere between the Qantas club and the boarding gate, I leapt upon a copy of *Freakonomics*. I devoured Steven Levitt and Stephen Dubner's hugely entertaining bestseller well before my plane made it across the Pacific. I was compelled and amazed at the insights it provided into everyday behavior, but what struck me most of all was the amazing similarity between behavioral economics and evolutionary biology.

Evolutionary biology has a long history of borrowing ideas and statistical approaches from economics, and the intellectual traffic has flowed both ways. But here I was reading a bestseller about the "hidden side of everything" that audaciously asked, and answered in entertaining and thought-provoking ways, the kinds of questions my biologist colleagues and I have been answering, using animals, for decades.

The sciences of evolution and economics are the two great frameworks within which we understand the rewards and costs that drive human behavior. Behavioral economists study how people respond to incentives by weighing them in terms of money or in other currencies like time, happiness, status, and that vague but all-important catch-all quantity called utility. This is where evolution comes in. Evolutionary biologists ask why people (and other organisms) respond to certain incentives rather than to others. Why do people like to have money? Why do we gamble and buy stuff that we will never need? And why are happiness and status so important to most of us?

Economics can give answers to these *why* questions. (We like money because we can buy stuff, we like stuff because buying it makes us happy.) But every question answered in this way opens up another; only by understanding evolution can we answer the big underlying *WHY?* It is because utility, the economic concept that most closely translates to "the thing that an individual wants to maximize," is given real meaning by evolution. Deep down people want happiness, wealth, security, warmth, well-being, nutritious food, status and good sex because these are the rewards natural selection has shaped for us. These rewards remind and encourage us to do the things that maximize evolutionary fitness. Or that would have maximized fitness in the environment of our ancestors.

While evolution and economics are cozying up (Michael Shermer calls this new synthesis "Evonomics"), the relationship between evolutionary and cultural explanations of human behavior is frosty. Part of the reason is in the way we explain our behavior; distinguishing the effects of genes and environment or the effects of biology and culture as if they were competing alternatives. It is not uncommon to hear about behaviors that are "in our genes," or "entirely cultural," as though the effects of genes and culture could be neatly partitioned.

Our most common metaphor for understanding human behavior is the computer, where biology is the hardware and culture the software. How often have you read about behaviors that are "wired," "hardwired" or "programmed"? This view causes all sorts of trouble. For a start, if we think of our evolved biology as the hardware, then it is too easy to think that our biology is deterministic and unwavering. And if something is "in our genes," then who are we to defy it?

Some social scientists still dogmatically assert that the environment, learning and other social processes are by far the most important—or even the only—determinants of human behavior. And some reject biological explanation in any form. To them, it is more comfortable to assume we all got roughly the same standard-issue computer, and that every difference between individual people is due to differences in how our experiences have programmed that machine. A computer can easily be reprogrammed, but it is impossible to change the hardware of the human brain.

Or is it? We learn and develop precisely because our brains change with experience—most obviously by making and reinforcing connections (not "rewiring") between brain cells. Brains are, after all, physical organs. If you accept, as modern neuroscience shows us, that human thought and consciousness have material causes within the nerve tissue of the brain, then you must reject any distinction between our evolved biology and the effects of experience, culture and learning. Biology is the material basis of culture.

In his exceptional book *The Blank Slate: The Modern Denial of Human Nature,* Harvard psychologist Stephen Pinker comprehensively dismantles these false dichotomies between nature and nurture, genes and environment, biology and culture. These, he argues, are "the last wall standing in the landscape of knowledge." In order to benefit from this book, you will need to come with me beyond the last wall, rejecting the tendency to see genetics and evolution as deterministic and inflexible, and culture and environment as infinitely malleable.

The last wall still stands for many media outlets, who love nothing more than a two-horse race. Stories still follow the hackneyed formula of pitting nature against nurture, genes against environment and evolution against culture. Many stories trumpet the discovery of single important genes: *the* heart disease gene, *the* obesity gene, *the* gay gene or *the* infidelity gene. These discoveries are usually of a gene that changes by a few percent—and usually under only certain conditions—the probability of having heart disease, becoming obese, being gay or cheating on your spouse. Yet every one of the tens of thousands of genes in our genome influences many very different traits and thus several facets of our lives. Any given trait is the result of an enormous number of interactions between genes from all over the genome. Yet we are still wallowing in "*the (insert trait here) gene*" stories that reinforce the antiquated and harmful link between simple genetic inheritance and destiny.

Most traits I talk about in this book develop in response to instructions from hundreds or even thousands of genes from all over the genome, and each gene has miniscule effects. To understand these stories, you don't need to know the nitty-gritty detail of which genes are involved or how they work. There is only one piece of simple genetic inheritance that I will consider in this book, and the natural history of this gene eloquently illustrates how complex and politically fraught the effects of a single gene can be and how silly it is to view genes as deterministic.

There is a gene called *SRY*, which is carried by about half of all people. It massively increases the risk of the carrier being a murderer or a murder victim, ending up in jail, dying in an accident or succumbing to a range of diseases in old age. But the effects of *SRY* are not all bad: carriers have, on average, exactly the same evolutionary fitness as non-carriers. Even though many more carriers than non-carriers of *SRY* never have any offspring at all, the biggest winners in evolutionary history, those people who have left the most descendants, have all been carriers.

SRY is the sex-determining region on the Y chromosome. It encodes the crucial genetic instruction that, via a complex molecular chain reaction, eventually turns the developing embryo into a male. Without this instruction, the body follows the default settings and develops into a female. Only men and boys have a copy of *SRY*, and it is fair to say that the effects of *SRY* are as deterministic as it gets—pretty much every embryo with an intact *SRY* ends up biologically male. But *SRY* does not do everything that is required to make a male body; it merely signals to other genes to prepare the body for the challenges of maleness. In the absence of the signal from *SRY*, the body prepares for the challenges of femaleness. How each gene goes about its job depends on the strength and timing of thousands of other signals. There are genes involved in the making and receiving of every one of those signals, and the action of those genes and the effectiveness of those signals is influenced at every step by the environment in which those genes find themselves.

While *SRY* determines maleness, the end product is extremely variable. Think of how the various men you know differ in muscle bulk, hairiness, voice pitch, aggression, or any of the other attributes that tend to be associated with masculinity. Similarly individual women vary just as much in these same traits, usually within a range that is different from but overlaps the range for men. Just as *SRY* and, therefore, maleness increases the chance of somebody being tall, muscular and hairy, it also increases the risk of a person being homicidal. But being a man does not make somebody a murderer, and it does not mean that women cannot be murderers either. Nothing is determined.

Not even maleness or femaleness is destiny. Men and women vary enormously, not only on a gradient from Barbie-doll hyper-femininity to GI Joe hyper-masculinity, but also in having different combinations of traits that we typically think of as masculine or feminine. So much so that some people with very typical bodies and behaviors for their biological sex self-identify as the other gender.

The differences between men and women and the complexities of gender identity are areas where evolutionary biologists often fear to tread, reluctant to risk lending even slight legitimacy to sexism, marginalization or oppression. Yet it is clear that the simple developmental switch that made each of us either a male or a female has long-lasting and often broadly predictable effects on our biology. I am especially concerned in the coming chapters with understanding how biological differences between men and women arise, how these differences affect the lives we lead, and how biological differences interact with culture and economics to shape our lives and the power relations between women and men.

Is, ought and wishful thinking

Creationists and religious fundamentalists despise evolutionary biology because it asks questions about who we are and how we live our lives. Creation myths are parables, written not only to explain to a curious flock how things came to be the way they are, but also to impose on a pliable flock an idea of how they ought to be. Science is a very different way of explaining the world. Scientists ask questions because they genuinely want to know the answer. And they expect that sometimes the answer they get will not be a comfortable or a comforting one.

The late paleontologist Stephen Jay Gould, giant among popular authors on evolution, was one of my early intellectual heroes. In 1979, he and his geneticist colleague Dick Lewontin viciously scorned the evolutionary study of human behavior (then called sociobiology) for the high crimes of "just-so storytelling" and "Panglossism." Just-so storytellers, they argued, dream up ideas about how some or other feature of an organism might have evolved and present it as a scientific account. Such accounts are no better than Rudyard Kipling's *Just So Stories:* entertaining children's nonsense tales. In Kipling's "The Elephant's Child," for example, the elephant's trunk originated from an epic tug-of-war between the insatiably curious Elephant's Child and a crocodile that bit his nose on "the banks of the great grey-green, greasy Limpopo River, all set about with fever-trees."

Panglossism is the optimistic worldview that everything that happens is for the best. Voltaire's Dr. Pangloss taught his young pupil, Candide, that they lived in the best of all possible worlds. In this world noses were designed to wear spectacles, legs to wear britches, and even Pangloss' syphilis was part of a benign grand plan. Panglossism is a pompous name for the naturalistic fallacy—inferring the way that things should be, and what is right, from the way that things are. This fallacy is the kind of lazy and unimaginative thinking that is often used to legitimize the status quo as the natural—or even divine—world order.

Like Gould and Lewontin, I too believe that we should not infer how things should be from how they are. Evolutionary accounts of human behavior attempt to understand all that is good in human nature, including cooperation, altruism, charity, love, kindness and modesty. Yet those virtues seldom evoke as much popular interest as the homicide, siblicide, infanticide, violence, rape, slavery, promiscuity, cannibalism, cheating, revenge and punishment that we also study. People do all these wonderful and heinous things and have done them for as long as there have been people.

Gould and Lewontin's Marxist acolytes and those dyed-in-the-wool social constructionists who still wield "Panglossism" and "just-so stories" as pejoratives would probably squirm at the realization that they share the same underlying fear about evolution. They shudder alike at the idea that the hideous and demonic sides of humanity might be adaptive because they fail to see the difference between understanding the dark aspects of human nature and justifying or even excusing them. It is far more convenient to dismiss our most appalling characteristics as evil, or to blame faceless societal bogeymen like "The Media," "The Patriarchy" or "Class Conflict," than it is to own up to their true origin.

But bogeymen are just a symptom of laziness. If the media are a tool of the oppressor we must ask why the oppressor needs to wield them. If the patriarchy keeps women down we should be asking why there is a patriarchy at all and to whose benefit it acts. The answers are not always obvious or even simple. At times in this book I will consider how social power and conflict arise from the different evolutionary interests of individuals, and hopefully this will reinforce why evolution is the missing piece of the puzzle that is human behavior. A mature understanding of evolution can help us to understand every virtue and every sin; not only why normal people do both heinous and virtuous things, but why we find those things heinous or virtuous in the first place. In chapter 8, for example, I will consider why, in many parts of the world, female fetuses are far more likely to be aborted, and newborn girls killed. But just because the origin of female feticide and infanticide can be understood with some recourse to evolution does not mean that by understanding these hideous crimes we should find any single case even slightly more acceptable.

I caution against the naturalistic fallacy because every day the idea that some behaviors are natural is used by somebody, often speciously, to legitimize their chosen way of life. Church leaders and conservative politicians often claim that lifelong heterosexual monogamy is the only natural way. Other people are equally vehement that we are naturally promiscuous. Others, still, assert that homosexual love is unnatural, and their opponents argue it is every bit as natural as the heterosexual form. Evolution can help us understand our behavior, but evolutionary history seldom yields simple moral ordinations to questions as complex and fraught as promiscuity or sexuality. First we should aim to understand how we came to be as we are, and the broad range of possible behaviors that are part of our natural human capacity. Only with wisdom and care should we feed these ideas into the broader processes that our societies use to sift wrong from right.

In defense of just-so stories

Unlike Panglossism, the time has come to reclaim just-so storytelling from the pejorative sense that Gould and Lewontin gave it. When I was a boy, I loved my mother's stories about the habits of the animals we encountered on holidays to South Africa's Kruger National Park. Some of those stories have since been debunked by scientific evidence, but this illustrates the value of just-so stories. It is possible to come up with imaginative stories about who we are or where we came from, but the thing that makes science special and sets it apart from other types of storytelling is that scientific narratives are not private truths but testable hypotheses. Because science is, above all, a method for sorting and separating ideas, there is nothing wrong with generating an incorrect hypothesis as long as it is then tested using the scientific method. Bad ideas will be rejected; perhaps not immediately, but eventually.

I want, in the next 11 chapters, to convey the true power of evolutionary biology to help us understand who we are, especially when evolution takes its place alongside economics and the scientific study of culture. In order to do so I need occasionally to flirt with just-so storytelling. Most of the ideas in this book are supported by good evidence, others only with preliminary evidence, and yet others are embryonic hypotheses. I am sure many of these ideas will turn out to be right and others will in time be refuted, but having the ideas in the first place is an important step without which science cannot make progress. As the pioneering ecologist Robert MacArthur said, "There are many things worse than being wrong. One is to be trivial." So the approach I take in this book, and one I hope other authors will take too, is to write about human evolution in a way that does not unduly fear being wrong if it means we can transcend the trivial.

1

The weight of our ancestry

Many parts of the modern world, including the wealthiest industrialized nations, face an enormous crisis of obesity. Part of the reason is that people are adapted to the diets of our hunter-gatherer and subsistence-farmer ancestors, for whom starvation was always a bigger problem than being overweight.

AS A CURIOUS YOUNG MAN at Cambridge, years before he set out on his round-the-world voyage and decades before he published the *Origin of Species*, Charles Darwin was a member of a strange group. The Glutton Club were students who met weekly, bravely dining on "birds and beasts which were before unknown to human palate," such as hawk and bittern. The group abandoned their culinary adventuring after an attempt to eat an old brown owl, which, Darwin later said, was "indescribable." Adventurous eaters as Darwin and his privileged fellow gluttons were, they would find the diets we eat in the twenty-first century quite bizarre.

Imagine if Darwin were to visit me in my office today. He and I, having caught up on all the major developments in evolutionary biology since he died in 1882, might fall to chatting about the eating habits of Victorian gentlemen, and of modern-day scientists. To satisfy his insatiable curiosity I might recount what I ate yesterday: ruby grapefruit and a boiled egg with toast for breakfast; apple mid-morning; beef vindaloo with rice for lunch; macadamia nuts and almonds mid-afternoon; a pasta dish with broccoli, anchovies, garlic and olives for dinner; and strawberries and dark chocolate for dessert. Apart from understandable revulsion at my strange eating habits, he would count eight different fruits, nuts or vegetables, three cereal products, two meats, an egg and some chocolate.

Darwin might ask how I could afford to eat like this on an ordinary day? Where did my servant get all the ingredients? How is it possible for broccoli, strawberries, grapefruit and apples to be in season at the same time in mid-winter Sydney? Although Darwin would be quite at home with the discussion on evolution, and overjoyed at developments in genetics, the bit about food would probably make him want to skip dinner. And Darwin was a wealthy gentleman who was still alive just over a century ago. What would the great scientists and mathematicians Isaac Newton (died 1727), Muhammad ibn Mūsā al-Khwārizmī (died 850) or Archimedes of Syracuse (died 212 BC) have made of my modern everyday feast? And what would an ancient hunter-gatherer ancestor have thought?

Some time in the first ten years of this century, the number of people who are dangerously overweight exceeded, for the first time ever, the number of people who are undernourished. According to the World Health Organization about 600 million adults worldwide are obese, and 2 billion are overweight. Every year a greater proportion of people in a larger number of countries are becoming overweight or obese, and it is happening at earlier ages. From the rise in obesity flows a cascade of health and social problems, the very real threat that life expectancy will decline for the first time in centuries, and the twin burdens of higher healthcare costs and lower workforce productivity. In the USA at least, fat is the new smoking. Books and movies like Erik Schlosser's *Fast Food Nation* and Morgan Spurlock's *Supersize Me* challenge mainstream culture. Food activists periodically declare War on Fat, with school canteens and fast-food chains the highest profile battlegrounds.

Waistlines have been ballooning for more than a generation in most developed countries and a similar increase is beginning to bite in the developing world. As a rough guide, people are said to be overweight when their body mass index (BMI, which is weight, in kilograms, divided by the square of height, in meters) exceeds 25, and they are officially obese when their BMI exceeds 30. Converting to pounds and inches, a man like me who stands 5 feet 10 inches tall is overweight when he gets above 175 pounds and obese when he hits 210 pounds. A woman who is 5 feet 4 inches tall, about average for an American woman, is overweight beyond 145 pounds and obese at 175 pounds. These are only rough guides. For example, BMI does not take muscle mass into account. Those obnoxiously buffed folks down at the gym often have BMIs above 25 but have so little body fat that their abdominal muscles take on that freakish and highly unnatural six-pack look.

Nonetheless if you aren't some kind of hyperactive gym bunny, and if you don't mind a bit of food and a regular beer or wine, then you are probably going to have to stare down a number close to 25 at some time in your life. The health consequences of weighing too much are detectable once you join these ranks and they escalate as you get fatter. There are massive increases in type-2 diabetes, high blood pressure, stroke, coronary heart disease and polycystic ovary syndrome, but there are also mental and physical health costs of the social exclusion and low self-esteem that often come with being obese.

The obesity crisis hit in the blink of an eye—a few generations at the most—and it isn't going away soon. It is impossible to ignore the fact that social, cultural, environmental and economic explanations are going to be important if we are to understand obesity. Some researchers use the fact that the obesity crisis emerged so recently as a reason to reject or ignore evolutionary or biological explanations for the crisis. After all, they argue, evolution takes thousands of generations. While they are certainly correct that we are not evolving to become fatter, obesity certainly has deep evolutionary roots among its causes. In fact, obesity illustrates as well as any other problem how tightly the effects of our evolved genes and our environment intertwine.

The social and economic triggers of obesity originate in our evolutionary past. The combinations of genes that we inherited from our ancestors are adapted to the environments where those ancestors lived and bred, not the environments we inhabit today. Those genes are by no means inflexible to circumstance—we have genetic information that tells our bodies what to do in a famine and what to do in times of plenty, what to do when we aren't getting enough protein and what to do when we need more salt.

Nonetheless natural selection wrote this genetic instruction manual over thousands of generations when life was very different for most people. Imagine trying to assemble a desk that you bought from IKEA. But only when you get home and open the box you find the instructions for a *bookcase* in the same product range (you have to imagine this because I am sure the fastidious folks at IKEA have *never* made this kind of mistake). Those instructions will be more useful than having nothing at all, but there are going to be some steps in the construction for which the instructions are of little or no help. That is where we and our bodies find ourselves today: with genetic instructions that don't always suit our circumstances.

Obesity is a biological problem because our biological bodies are laying down dangerous amounts of fat. It is an evolutionary problem because the way our bodies react to modern environments can best be understood in terms of how they evolved to function and especially to handle energy and nutrients in ancient environments. Weight gain results from a complex set of interactions between our evolutionary past and our environmental present, and if we are to understand the problem we need to meet it at the crossroads. In this chapter and the next, I will explore why people in many parts of the world are experiencing such spectacular and catastrophic weight gain. In so doing I will illustrate the usefulness of an evolutionary perspective. But I will also show that evolution and biology complement rather than compete with an understanding of the complex environmental, economic, social and cultural factors behind the obesity crisis.

Energy hoarders

The basic equation for weight gain or loss is straightforward. Take the total energy value of the foods you consume in a day, then subtract the energy your body uses in exercise and the small amount of energy in the matter that you flush down the toilet. The number you have left is the energy surplus or deficit, depending on whether you have eaten more or fewer kilocalories than you used up (a kilocalorie is the same as the calorie on a US nutrition facts label). Energy surpluses are saved by the body in the form of fat, and deficits require you to draw down on your fatty savings. All of this oversimplifies some fascinating and complicated physiology, but in essence weight gain or loss boils down to the difference between energy intake and output.

The energy budget is not just some form of passive balance sheet, however. In fact it is managed by the body far better than even the way the most retentive accountant manages a financial budget. What and how much we eat, the amount of energy we burn off and how and where we store excess energy as fat are all regulated by the body via a sophisticated system of behaviors, nerves and hormones that natural selection has refined over millennia.

Like all animals, people have always faced two basic problems when it comes to dietary energy: having too little or having too much. On the face of it, the curvaceous Mae West may have been right: too much of a good thing can indeed be wonderful. The energy we take in is used by the body to fuel the working of our organs, our exercise and our work, as well as to keep our bodies warm. Energy from carbohydrates is easiest for our bodies to use, but fats—almost twice as dense with energy as carbohydrates—are very good fuel too. Protein contains about the same energy per gram of food as carbohydrates, but the energy in proteins is hardest to break down and we need the amino acids to build muscles and other tissues, so protein is used last. If we find ourselves taking in too much energy for a time, the liver can turn it into fat where it can be stored in all the usual inconvenient places, or we can burn some of the excess off.

Natural selection shapes the rules for how animal bodies deal with excess energy, and there are some radical differences between species. Mice, for example, cannot afford to carry extra fat around and become a bigger and slower-moving target for predatory cats, birds and snakes. As a result, mice burn off about 90 percent of the excess energy by generating extra heat and automatically doing more exercise. For humans, however, this figure is a rather dismal 25 percent because we have evolved to be frugal with our energy reserves. How different for us would the current crisis of overnutrition be if we were more like mice—at least in our ability to burn off excess energy.

Dietary energy is unlike money in one important respect: it is okay to save some for later, but there is a limit on how much we can store before too much of a good thing becomes a catastrophic thing. Unlike money in the bank, we have to carry our reserves around with us. This gets impractical and can be downright deadly. Many of the diseases associated with obesity involve high blood glucose levels and the hormone insulin. Insulin regulates glucose and fat metabolism and influences growth, weight and fertility. When somebody becomes obese, the cells where their fat is stored, and their muscles and liver, grow resistant to insulin. These tissues become less effective at taking up glucose from the blood or at breaking down fat to use as energy. It then becomes increasingly difficult to break down body fat, and people with insulin resistance can find it much harder to lose weight than other people.

In today's world, children and people in their reproductive prime are dying from the so-called "lifestyle diseases" that come with excessive weight. On top of the risks of dying, polycystic ovary syndrome, a condition associated with insulin resistance and obesity, is a leading cause of female infertility. Because evolution strongly favors traits that lead to reproductive success—even at the expense of life span—the reproductive problems experienced by the overweight and obese were selected against in the past. In fact, very strong selection on female fertility may be driving adaptation to modern diets as well, given the high rates of polycystic ovary syndrome and infertility in some ethnic groups such as South Indian migrants to the United Kingdom, who are currently experiencing explosive growth in obesity.

But for most of our evolutionary past, starvation was a far more common problem than being overweight ever was. To understand why most people find it easier to pack on weight than to shed it, it is worthwhile remembering that our ancestors always—starting long before even the first mammals—faced a very real danger of running out of fuel. Having too much fuel on board was far less often and far less severe a problem. Even today, 1.1 billion people—one of every six people currently alive—do not obtain enough energy from their diet to meet their biological needs. In 2001, Professor Jean Ziegler reported to the United Nations that 58 percent of the people who die worldwide each year die "directly or indirectly as a result of nutritional deficiencies, infections, epidemics or diseases which attack the body when its resistance and immunity have been weakened by undernourishment and hunger."

In evolutionary terms, the costs of starvation are absolute. There is a palpable risk of running out of fuel for the body's basic metabolic needs: keeping the body temperature where it needs to be, the brain working, the liver functioning, the kidneys excreting and the digestive system ready to extract the goodness from any morsel that comes our way. In addition, starvation and its equally ugly twin, malnutrition, increase susceptibility to parasites and diseases. Diarrhea, irritant of Western travelers and parents with children in daycare, becomes a ruthless killer of undernourished kids.

As if the risk of dying were not bad enough, undernourishment in the womb or during childhood can stunt growth, delay the onset of the menstrual cycle, bring forward menopause, or inhibit ovulation. Undernourished men have fewer, flimsier sperm than well-fed men. Those babies that underfed mothers do conceive are less likely to live to term, less likely to thrive and grow, and don't get as much care as babies of well-nourished mums. All of these symptoms of starvation impose the ultimate evolutionary price because starved people cannot reproduce as successfully as well-nourished people and thus they leave fewer descendants.

Many more of our ancestors starved than ever became obese. In every generation the people who avoided starvation or who coped best with the symptoms left more children and grandchildren than their friends, neighbors and enemies who did not avoid or cope with starvation. These ancestors left us the genes that helped them avoid or deal with starvation, but this may turn out to have been a dubious bequest. When we don't take in enough energy for several days, we have to draw down on our savings by metabolizing our fat reserves and then, if we run out of fat, our muscle protein, to release energy. Just like a diligent accountant, our bodies exercise prudence and thrift when we start to draw on our reserves. We become lethargic and less likely to exercise. Our bodies also use less energy to generate heat and to maintain our internal organs. The small intestine shrinks. Women stop ovulating and menstruating.

The hormone leptin, which is produced by our fat cells and which drops in concentration as we start to starve, is involved in many of these processes. Leptin, a powerful regulator of appetite, responds far more urgently to undereating than to overeating. As a general rule, our bodies tend to treat a shortfall in energy or nutrients as far more of an emergency than they do an excess: we get very hungry and our bodies go into thrift mode as soon as there is even a chance we will starve, but any excess energy is shrewdly squirrelled away against possible future starvation. The thrifty response that our bodies take to starvation, at least partially via the regulatory effects of leptin, explains, in sound evolutionary terms, why starvation diets do not often work, and why they so often implode in a fit of binge eating.

Spending energy in the modern world

There is a tussle between those who wish to blame the obesity crisis on gluttony and those who would blame it on sloth. As I hope this book will show, I am skeptical about apportioning blame at all, let alone deciding which deadly sin is involved. The weight of evidence in obesity research suggests that increased energy intake is responsible for much of the obesity crisis, but for individuals the battle to control body weight is a battle to balance energy intake with energy expenditure. And this means, at some point, exercise. I don't necessarily mean working the treadmill—or that most bizarre of modern inventions, the elliptical trainer—but rather the old-fashioned exercise that comes with work and with walking.

For a great many people, "work" now involves sitting down in an office, which we reach by car, bus or train. When we want water, we drink it from a tap or a bottle. When we need food we go to the refrigerator or a nearby restaurant. And when we come home, tired, from a day of "work" we lie in front of the television or read books on evolution. For our children, "play" can involve the twenty-first century's most popular imaginary friends: PlayStation, Wii, Nintendo, and the Internet. I am neither Luddite nor technophobe, and I don't think that any of these developments are necessarily a bad thing. But the next time you are watching television, driving to your local shops, or asking your five-year-old to explain again how to use her Nintendo DS, consider what somebody in Cambodia, India or Ethiopia might be doing.

A great many people in countries like these experience starvation or malnutrition at least once. And earning their meals mostly requires exhausting labor and walking distances that few of us would contemplate. In these countries, water often comes from miles away, and it must be fetched. Those who carry water know that it always weighs one kilogram per liter (or about 8.3 pounds per gallon). Fuel for cooking and heating comes from animal dung or wood, and it must be found, chopped or broken and then carried home. Work involves tending to subsistence crops, looking after domestic animals, fishing, or working in labor-intensive industries. Even those who don't labor physically burn more energy in these countries. Information technology workers in Bangalore, bureaucrats in Jakarta or office workers in Nairobi walk a great deal more in their daily commute than do their counterparts in San Jose, Brisbane or Manchester. In developing countries, wealthier schoolchildren who no longer walk to school are among the first groups to show signs of obesity.

If you have ever tried to monitor your exercise using a pedometer to count the steps you take, you may often need to make a special effort to reach the target of 10,000 steps per day that sets the benchmark for an "active" lifestyle. But how much of your energy budget is that? For somebody of my physique, walking 10,000 steps burns approximately 360 kilocalories. That is only 15 percent of my daily energy expenditure. Where does all of the other energy get burned? Most of it is burned on the basics of keeping us alive. Our brains are expensive hardware to run, burning roughly one-fifth of the energy used by the body. And our other organ systems are also very demanding. We also burn a lot of energy just keeping warm. Humans keep their body temperature very close to 37 degrees Celsius (98.6 degrees Fahrenheit). When we are cold, we urgently seek out warmth and insulation, and mechanisms far beneath our conscious awareness start burning more energy in order to increase the production of heat. As humans have mastered the production of adequate clothing, warm bedding, and well-heated, draft-proof housing, and as we have moved to a more sedentary, indoor lifestyle, so we have been liberated from the need to produce heat for regulating our body temperature.

There is little doubt that modern lifestyles lead to lower energy use than the lifestyles of our hunter-gatherer and farmer ancestors, mainly because we don't rely much on our own metabolic heat-production to stay warm, and we use less energy as we go about our daily lives. It doesn't take an evolutionary biologist to recognize that modern lifestyles can be far too sedentary, warm and comfortable for our own good. But we can learn a great deal more about how to turn the balance between our energy expenditure and intake to our advantage by considering the food we eat.

Leaving vegetarianism behind

Our ancestors have always lived in changing and variable environments. Many times over, the local climate cooled, warmed, became wetter or drier and often new species of plants and animals became established in the area. Each time people moved, sometimes only a few tens of miles, they faced quite different environments. Each of these changes would have altered the combinations of foods that our ancestors had available to them. Over many generations, changes in food availability and nutritional needs drive the evolution of genetic changes in the ways our bodies handle energy and nutrients (metabolism), as well as the sensory and psychological pathways that stimulate us to eat or stop eating and guide us in what we choose to eat. But the evolution of diet is more than a duet between environmental change and natural selection. Culture, the transmission of learning from one individual to another, is an equal third player both responding to and driving environmental and genetic changes.

Human societies have accumulated colossal bodies of learned knowledge about where and when to find the best foods, how to dig them up, catch, kill or cultivate them and how to prepare and eat them.

Our prodigious capacity for learning and for communication means that societies accumulate cultural knowledge about food, enabling each society to better exploit the range of available foods, to expand into and use new habitats with new foods, and to exploit one another's labor in the production of foods. At the same time, our bodies and genes have continued to evolve in step with our cultures.

Genes, culture and the food environment are continually shaping one another, but there are three important transitions in the evolution of the human diet that have completely reshaped the way we eat: the transitions from vegetarian ape to human hunter-gatherer, from hunter-gatherer to farmer and from farmer to manufacturer. Each of these transitions can teach us something about the deep causes of the modern obesity crisis.

We all descend from a long line of vegetarians. Chimpanzees, bonobos, gorillas and orangutans obtain the vast majority—up to 99 percent—of their nutrients from plant materials. They occasionally supplement these leaves, fruits, seeds and roots by eating protein-rich animal foods like termites, grubs and scavenged or hunted vertebrate carcasses. Most of the plants they eat contain very little energy, and apes guzzle with gluttonous enthusiasm any energy-rich fruits they are lucky enough to find. This has led to the strong impression that foraging primates have evolved an overwhelmingly strong drive to maximize their energy intake. It is a drive that our ancestors probably shared.

Between 5 million years ago, when we shared an ancestor with modern chimpanzees, and 2 million years ago, our ancestors gradually evolved from vegetarian apes into upright-walking hunter-gatherers. This new ancestor, *Homo erectus,* was taller, heavier and with a bigger brain than any hominid that had come before. Bigger bodies and bigger brains require more energy, and these changes evolved together with a tendency to supplement the vegetarian diet by hunting for animals. Hunting provided the valuable protein and energy-dense fat that allowed

Homo erectus to grow large. And larger hunters with bigger brains were better able to outsmart and overpower their prey.

The upright stance also freed the hands to make and use tools. Bigger brains would have led to better tools, more successful hunting, and cooperation to gather root and tuber foods. When *Homo erectus* got the hang of cooking with fire, a whole world of culinary possibilities opened up. Cooking helps to break food down, making both vegetables and meat more easily digestible. When we cook our food we have to spend much less energy chewing and then digesting the food and we are able to absorb more of the energy and protein in the food.

For at least 2 million years, the ancestors of everybody alive today lived a hunter-gatherer existence. That makes about 90,000 generations of adaptation to an increasingly human hunter-gatherer diet, in which natural selection optimized how our bodies use the food and store the energy from the animals that hunters and fishers bring in and the plants and small animals that our ancestors gathered. But it is a mistake to think that all hunter-gatherers were the same. As our ancestors spread out across Africa and then across the world, they co-evolved with their local diets. Local differences in the species of food plants and animals that were available and cultural differences that arose between groups from this time onward mean that there was no single hunter-gatherer diet among the ancestors of all modern people. Hunter-gatherers in equatorial Africa would have eaten nuts, roots and the lean meat of mammals; those living in Arctic Canada got most of their energy from oily fish and whale meat and blubber. But however different the diets of these ancestors were, they were all living off plant matter and animals harvested from the wild.

Modern humans get much more of their energy from carbohydrates and less from proteins or fats than our hunter-gatherer ancestors ever did. Recent estimates suggest that most hunter-gatherers got about 35 percent of their energy from carbohydrates, whereas modern Western diets range from about 50 to as high as 65 percent carbohydrates. It matters, too, where the carbohydrate energy comes from. Hunter-gatherers get most of their carbohydrate energy from fruits, tubers and other vegetables that are three to four times higher in undigestible dietary fiber than the domesticated vegetables we buy in modern supermarkets. Fibrous foods take a lot of effort and energy to digest, and energy becomes available gradually because complex starch molecules break down slowly into sugars. In nutritional jargon, the plants eaten by our hunter-gatherer ancestors were the ultimate low–glycemic index (GI) foods. Modern Americans get 15 percent of their energy from sugars, and most of the rest from easily digested simple starches in refined cereal grains, whereas hunter-gatherers get no more than 2 percent of their energy from sugars (in the form of honey), and have no cereal grains.

For most of human history, our ancestors living in equatorial Africa probably got close to 35 percent of their energy from fat. Although modern humans get less of their total energy from fat, we eat more saturated fat—the least healthy kind. The drop in modern fat consumption, particularly the healthier fats, is probably because a smaller proportion of our modern energy budget comes from meat and nuts than it did when our ancestors were hunter-gatherers. For the same reason we only get around 15 percent of our total dietary energy from protein, rather than the 30+ percent that hunter-gatherers get.

The worst mistake in history

Although a small and ever-shrinking number of hunter-gatherer societies persist in wild places like the Amazon and the Kalahari, most people alive today are descended from farmers. Agriculture emerged several times in many different parts of the world, including the Levant and Mesopotamia, China, Southeast Asia, Sahel Africa, the Americas and Papua. The very earliest farmers lived in the fertile crescent of Mesopotamia (modern Iraq) and the Levant (Syria, Jordan, Lebanon, Israel and the Palestinian territories) no earlier than 12,000 years ago.

Some other transitions to agriculture were much more recent; as late as 800–1000 AD in parts of North America.

Every time humans invented agriculture, they domesticated at least one major carbohydrate source: wheat and barley in the Levant, Mesopotamia, India and temperate Asia; millet and sorghum in Africa; rice in southern China; corn and potatoes in the Americas; and sugarcane in Papua. Other crops and the domestication of livestock often followed, but every time agriculture gained a toehold against the hunter-gatherer lifestyle it did so because it allowed farmers to cultivate and store large quantities of carbohydrate energy.

Each time agriculture arose, it took hundreds or even thousands of years for people to move from a nomadic hunting and gathering existence to living as sedentary farmers. One reason that agriculture took hold slowly is because crops evolved gradually. Favorite gathered plant foods, like the grassy ancestors of wheat, corn and rice, first grew accidentally around well-used hunter-gatherer camps. Seeds discarded on rubbish piles, or seeds that survived being eaten and passed through people's guts, germinated and flourished near well-used campsites. Mutations in those foods that made them easier to harvest and process, or that made the edible parts bigger and more nutritious, were favored, inadvertently at first but then increasingly as people began to master the art of deliberately selecting and replanting the seeds of the best individuals.

As some groups of people domesticated crops and then livestock, social structure shifted from small nomadic family groups to larger, more sedentary and often more permanent societies. The massive increase in relatively cheap energy from cereal grains meant that farmers often reached ten times the density that hunter-gatherers had achieved on the same lands, and farming groups tended to spread out to farm new country, usually at the expense of hunter-gatherers living on those lands.

Despite its historic significance, the transition to agriculture did not leave most people any better off by the time the transition was over than their ancestors had been as hunter-gatherers. In *An Edible History of Humanity*, Tom Standage says the invention of farming may have been "the worst mistake in the history of the human race." High population densities and new means of food production drove social changes on a scale not seen before or since. These included longer working hours, the formation of exploitative ruling elites and priesthoods, and the rise of infectious diseases like measles and smallpox. Women were temporarily freed from having to carry each baby until it could walk with the family or tribe, only to find themselves pregnant twice as often and with considerably less social power than they had as hunter-gatherers.

The food was ghastly too. Modern hunter-gatherer diets comprise more than a hundred commonly consumed food items, each with quite different compositions of protein, fats and carbohydrates as well as the dozens of other essential micronutrients. Our hunter-gatherer ancestors probably had a similarly varied diet. But each time people invented agriculture their diets became drastically narrower, dominated by their main cereal crops. Over tens of generations other foods probably fell out of the cultural memory. Wild animal populations dwindled in areas with large settlements because the extra mouths supported by cheap carbohydrate crops also demanded meat. The result was a shortage of protein and of micronutrients for all but the wealthiest and most powerful people. As agriculture developed, so farmers domesticated new crops providing a more varied range of nutrients, but most people could not afford a varied diet.

Although farming gave our ancestors more access to carbohydrate energy, human populations expanded to use up this extra energy. As a result most individuals were soon no better off energetically than their hunting and gathering ancestors had been. Although hunter-gatherers certainly experienced food shortages, their broad diets and nomadic lifestyles buffered them from long annual hungry seasons or anything as catastrophic as a crop failure. Subsistence farmers, however, suffer every year through lean months leading up to the harvest. Things get worse in years when crops fail, plagues strike and weather is poor, bringing famine. Our farming ancestors scraped the earth to meet their energy and protein needs, perennially at risk of starvation. During this period of human history natural selection strongly favored those metabolic tricks that encouraged people to overeat when high-quality carbohydrates and fats were available and store them as fat for the lean period that would inevitably follow.

Even when there was enough to eat, the dependence on carbohydrate crops caused severe health problems. Paleopathologists studying skeletons and teeth from ancient peoples before and during transitions to agriculture note that the hunter-gatherers were almost always tall and healthy with relatively good teeth. But once societies made the transition to agriculture, children grew more slowly and ended up as smaller adults. Iron-deficiency and anemia were common, as were deficiencies in several amino acids and vitamins that can only be acquired from the diet, and bones were weak. Dentists and farmers are among the modern-day professions that can be grateful for the evolution of agriculture. The sudden increase in dietary carbohydrates appears to have resulted in a threefold increase in dental cavities as a result of the acids produced by bacteria digesting sugars in the mouth.

Although our ancestors hunted and gathered for a hundred times more of our evolutionary past than they have farmed, it is misguided to imagine our bodies are stuck entirely in the pre-agricultural Pleistocene epoch as some fad diets claim. Most of our ancestors spent the recent evolutionary past in subsistence farming economies. According to evolutionary genetics, even a few recent generations of strong natural selection can cause plenty of evolutionary change. Experiments show that caterpillars adapt within fewer than ten generations to a high carbohydrate diet, increasing their capacity to burn off excess energy and not become obese. There is no reason to believe that humans have not evolved in response to the massive changes in diet that came with agriculture. This evolution occurred when natural selection weeded out the genes of the most anemic, toothless and weak in favor of the genes that made people better able to handle the new conditions.

From these observations comes a prediction—the raw material for scientific investigation. My prediction is that agriculture has left a signature on the diet choices and metabolisms of people whose ancestors have farmed for many generations. A long history of agriculture will have equipped some groups of modern people for a diet high in starch cereal grains with substantial food shortages occurring annually and catastrophic shortages occurring more than once a generation. Modern hunter-gatherers and groups that have had a shorter period co-evolving with agriculture should be less well adapted to a high carbohydrate diet and will be truly stuck in the Pleistocene. In the next chapter I will consider this possibility, but for now I need to introduce the last major transition in the human diet.

Industrialized carbohydrates

The Industrial Revolution and frantic technological innovation since that time have entirely reshaped how we farm, how food reaches us and how it is stored. In so doing it has dramatically changed what and how much we eat. The key transition here is from subsistence or small-scale farming to large-scale industrial farming with farmed commodities traded in bulk, transported globally and often purchased in large supermarkets or fast-food outlets. Like the transition to farming, this was a gradual transition. In some parts of the world, people have made part of this transition and in other parts most people still farm for their own subsistence.

Industrial innovations like mechanized farming, food processing, refrigeration, rapid global refrigerated transport, effective commodities market systems, sophisticated advertising and the rise of multinational supermarket and restaurant chains mean that people in industrialized nations as well as the wealthier people in developing nations have a diet that is broader, easier to obtain and more abundant than at any other time in human history. The dominant theme in the history of the human diet, from the advent of agriculture to the present day, is one of increasingly easy and cheap access to carbohydrate energy. While farming dramatically increased our ancestors' dependence on starchy carbohydrates, since the nineteenth century our intake of both starches and sugars has gone ballistic.

Sugarcane was domesticated in Papua and Southeast Asia, but spread to the Middle East and then throughout the tropics. The introduction of sugar to the New World rewrote human social and economic history. The sugar colonies of the Caribbean drove much of British and European industrialization. Their warm sunny climates and suitable rainfall made large-scale sugar growing possible, and a combination of slave labor and the world's first mechanized production lines made sugar cheaper than it had ever been before. Refined sugar suddenly changed from expensive delicacy to an affordable and important energy source for British and European workers. Molasses sent to New England was distilled into rum, much of which was sent to Africa in order to buy slaves who were sent to work 16-hour days in the sugar fields and refineries of the West Indies, fueled by the poorest quality salted-and-dried cod from New England and Nova Scotia. Thus sugar became an essential link between colonial governments, European manufacturers, slave traders, rum runners, and cod fishermen. It remains an important export for many Caribbean nations as well as Brazil and tropical countries of Africa, India and the Pacific. Brazil, the world's largest grower of sugarcane, also uses sugar-derived ethanol extensively as a biofuel. With the current energy crisis, fuel for bodies is increasingly a fuel for our machines too.

Sugar from tropical cane, and from corn as well as from beets and maple in more temperate climates, is more widely available and cheaper today than at any time in our history. Government price subsidies for sugar from otherwise uncompetitive farms in Europe, Japan and

America artificially inflate the worldwide supply of sugar, and thus distort the sugar price downward in a way that is unrivaled by any other commodity on the world market. Is it any surprise, then, that the proportion of dietary energy coming from sugar has increased so dramatically? Processed foods, fast foods, baked goods and nearly all beverages contain added sugar.

Carbohydrates other than sugar are also cheaper and more abundant than ever before. Twentieth-century mass production, refrigeration and fast-food delivery all led to cheaper, more widely available carbohydrates. For example, large food companies have industrialized potato farming and the manufacture and distribution of French fries to such an extent, creating such economies of scale, that it is almost trivially cheap for major fast-food chains to produce a serving of fries. French fries—irresistibly salty, fatty and packed with carbohydrate energy—are a license for fast-food companies to print money.

Out of the frying pan and into the fire

The abundance and variety of modern food is part of the problem driving the obesity crisis. Modern corpulence comes from an interaction between a genome fashioned first by hunting and gathering and then—for most people—by subsistence agriculture, with a modern environment that is unlike anything our genes have ever seen before. Food is cheaper, more abundant, more refined, sweeter and higher in saturated fat and artificial flavors than at any other time in history. It is also often prepared in bulk, stored and transported, marketed and sold to us in ways that save us a great deal of time. And it is ever more true that time is money.

Throughout evolutionary history, the people who ate the right foods, stored the right amounts of fat and burned off the right amounts of excess energy were also the people who survived famines and whose bodies grew strongest and healthiest without succumbing to obesity.

These are the people who left the most descendants and they are the people we count as our ancestors. As hunter-gatherers, our ancestors were mostly better than their contemporaries at gathering roots, berries and other plant-based foods and at hunting animals. They were the ones who were good at seeking out and securing the occasional fatty and sweet foods that delivered massive energy bonuses. And they were the ones whose bodies stored what they could whenever there was surplus energy. With ancestors like these, is it any surprise that many people find it difficult to control their intake of sweet and fatty foods? The Krispy Kreme Original Glazed doughnut may be the perfect imitation of our ancestors' rarest and most precious meal—now brought to you piping hot for a low low price.

We are adapted to the diet of our ancestors. Adaptive explanations are necessarily backward-looking like this, especially for modern humans, because natural selection is not a design process but rather a consequence of all the random little things that have influenced who lived and who died, who reproduced and who did not. This blind and inefficient process can make some colossal mistakes—or at least it can give rise to adaptations that turn out in hindsight to look like colossal mistakes—like our tendency to lay down excess calories as fat as if the next famine was just around the corner. Throughout history this was a good strategy because more often than not famine really *was* just around the next corner, but for people living in societies that have escaped from hunger and where the only thing lurking around the corner is a doughnut shop, it is a very real disaster.

2

Obesity is not for everyone

Evolution fits animals out to make the best of a range of conditions. We can understand why the obesity crisis is worse in some places than in others and why some people put on more weight than others by understanding how our evolved bodies respond under different economic and environmental conditions.

THE ISLAND OF KOSRAE, A speck of reef-fringed rainforest in the middle of the Pacific, is one of the most gorgeous and unspoiled places on earth. Tourism has not yet desecrated the island with four-story cabana-style resorts because Kosrae is so utterly remote from just about anywhere else. Planning a diving holiday some years ago, I was desperate to visit the ancient and spectacular reefs off Kosrae. But the ridiculous number of flights and connections I would have had to make meant that just getting there and back would have taken me longer than the amount of leave I had available. Such an idyllic and out-of-the-way island, replete with fresh reef fish, rainforest fruit and plenty of sunshine, should be the perfect place to live a healthy and well-nourished life. Unfortunately, at last estimate, nine out of ten adult residents of Kosrae were officially overweight and three out of five were obese.

Those are staggering numbers, but Kosrae is by no means the only place where such a high proportion of people are overweight. In fact the Federated States of Micronesia, of which Kosrae is the smallest state, ranks only sixth among nations by the proportion of people who are obese. Strikingly, the top seven are all Pacific islands. Obesity was not always this common in the Pacific; early European explorers found the inhabitants physically impressive, lean and powerful in build. But obesity erupted in the Pacific, as in many other parts of the world, in the late twentieth century. Luckily we know a lot about the social and economic changes that occurred around the same time, which are probably part of the cause.

Although the obesity crisis is touching every country in the world, there are enormous differences among and within countries in how the crisis is unfolding. A look at the national obesity league tables (see p. 40) turns up a few surprises. The seven nations with the highest levels of adult obesity are Pacific island states, and six of the next nine are in the Middle East. The United States, where they do everything bigger and better, including obesity, only weighs in at 10th on this list, with Australia in 20th and the United Kingdom in 21st place. That is not to underplay the scale of the problem in Australia, the United States or in any other nation. There is a global transition toward obesity that other nations have yet to follow. America and other developed countries are quite far down that road, but they have recently been dramatically overtaken by the Pacific island nations.

Among industrialized nations, it is a good time to be from Western Europe or Scandinavia, where fewer than 10 percent of people are obese. But that is still a pretty hefty one in ten adults. It is better still to be Japanese or South Korean where only 2 to 3 percent of adults are obese. However, the 20 nations with the very lowest percentages of obese citizens have largely undeveloped or developing economies with widespread poverty. Most of the 1.1 billion or so people who do not get enough to eat live in these countries, most of which are in Africa, the Indian subcontinent and the poorer parts of Asia.

The countries with the highest and lowest percentages of obese adults, according to data from the World Health Organization.

RANK	COUNTRY	ADULT OBESITY (%)
1	Nauru	78.7
2	Samoa	74.8
3	Tokelau	63.2
4	Kiribati	50.3
5	Marshall Islands	46.0
6	Federated States of Micronesia	44.0
7	French Polynesia	40.4
8	Saudi Arabia	36.1
9	Panama	33.9
10	United States	33.7
11	United Arab Emirates	32.8
12	Iraq	32.2
13	Mexico	29.4
14	Kuwait	29.0
15	Egypt	28.9
16	Bahrain	28.5
17	New Zealand	25.4
18	Macedonia	25.3
19	Seychelles	25.1
20	Australia	24.8
21	United Kingdom	24.0

RANK	COUNTRY	ADULT OBESITY (%)
118	Niger	3.2
119	Japan	3.1
120	Guinea	3.0
121	China	2.9
122	Togo	2.5
123	Burkina Faso	2.4
124	Malawi	2.4
125	Indonesia	2.4
126	Korea, South	2.4
127	India	2.1
128	Bangladesh	1.7
129	Chad	1.5
130	Rwanda	1.3
131	Cambodia	1.2
132	Laos	1.2
133	Central African Republic	1.1
134	Madagascar	1.0
135	Nepal	1.0
136	Ethiopia	0.7
137	Vietnam	0.4

Source: See www.robbrooks.net/rob-brooks/1317 for a full list of the 137 countries for which the WHO lists data.

In these developing countries, obesity plagues only the wealthy, just as it did everywhere until recently. The connection between wealth and obesity seems obvious: wealthy people can afford to eat well and don't have to labor hard to make a living or walk long distances to get necessities. As economies develop and people escape from poverty, the proportion of overweight and obese people increases. This explains why obesity rises with economic development and why wealthier countries tend to have higher incidences of obesity.

Puzzlingly, however, in wealthy industrialized nations like America, Britain and Australia, the relationship between wealth and obesity has flipped. In these places the poorest people living in the poorest regions are also the people most likely to be obese or overweight. Those people least likely to be in positions of economic or political power are also most at risk of obesity. In America, obese adults have had fewer years of formal education, earn lower incomes and are more likely to be from racial or ethnic minorities than adults who are of normal body weight. And within these groups, women are more likely to be obese than men. Indigenous peoples, another typically marginalized group, are finding obesity and the medical ills that accompany it to be the curse of modernization. Australian Aborigines and Native Americans, for example, are finding that diseases unknown by their parents and grandparents are suddenly rife in their communities. The same is true for Pacific islanders.

Gender differences and the privations of the poor, the indigenous and the working class are subjects often left well alone by serious evolutionary biologists who care deeply about sexism, alienation and poverty. As a result biologists have far too seldom and far too sporadically wielded the full force of evolutionary argument in seeking to understand and resolve many problems that our societies face. In this book I am trying to claim back some of the high ground; control of resources and the flow of energy are as much the domain of evolutionary biology as they are of economics or the social sciences. Differences among populations, among individuals, and between the sexes are the bread and butter of modern evolutionary biology. In this chapter I will show how geographic, socioeconomic and gender patterns in the unfolding global obesity crisis are part of a profound pattern at the confluence of social power, money, sex and more than 2 million years of evolutionary history.

The hungry and the choosy

Humans, like all animals, are constantly making choices about when to eat, what to eat, how much to eat and when to stop. By choice, I mean something more than conscious choice; anybody who has wallowed in engorged self-pity after a sumptuous dinner can testify that what we intend to do, what we know is good for us and what we eventually do are often very different things. A complex set of incentives and rewards govern our choices; these mechanisms have been shaped and reshaped by natural selection over our entire evolutionary history.

Why do some people eat so much? The first and most obvious of many suitable answers to this question is that food tastes good. We are eating more than ever because food tastes better than ever. Some people I know could make a strong case that the hot Krispy Kreme Original Glazed doughnut, the Big Mac or the pork dim sum are the high points of human cultural evolution. The fact that food tastes good is only a partial explanation for why we eat. Similar partial explanations for why people eat too much include: because they are hungry, because portion sizes are too large, because they are mysteriously compelled by food advertising, or because they only ever feel truly happy halfway through a quart of double chocolate ice cream. Explanations like these all fail to penetrate *why* food tastes good, *why* people get hungry and *why* we make pleasant associations with food.

These kinds of "Why?" questions are squarely on evolutionary turf: the realm of ultimate answers to big questions. In chapter 1, I argued that we have evolved to enjoy the foods that would have allowed our ancestors to reproduce and raise many offspring successfully. Natural selection shaped the mechanisms that make us feel hungry or satiated, the leptin signals and receptors that regulate our appetite, the ways in which our tissues respond to insulin in managing our energy budget, and the ways in which we learn about possible edible items using smell and the five tastes.

Five? When I was a boy and went to school I colored in pictures of the human tongue with areas for four tastes: sweet, sour, bitter and salty. Basically, the tongue is peppered with receptor cells for these four tastes, each stimulated by different groups of molecules or ions, although they are not really separated on the tongue like the map I colored in at school. But the Japanese have long recognized a fifth fundamental taste, which they call *umami*: a word meaning "tasty," "meaty" or "savory."

Japanese food scientist Kikunae Ikeda first isolated the chemical that gives seaweed broth its distinctive savory flavour in 1909, and helped it become the most successful food additive since salt and sugar: we know it as monosodium glutamate or MSG for short. Only in 2002 was the umami taste receptor identified, giving it the same status as the other four taste receptors I learned about in school. There are many other dimensions to taste, such as fatty, metallic and astringent (the dry taste of tannin-rich red wine or black tea), although the sensory mechanisms involved are mostly not yet resolved.

Protein-rich foods have strong umami flavours, and our ability to taste umami probably evolved to regulate our protein intake. Likewise sweetness is probably involved in identifying foods rich in carbohydrates, particularly sugars and simple starches. Although each food has a different blend of components and thus of flavors, there are very good reasons our taste receptors and our sense of smell detect the precise molecules that they do. These are the important molecules to recognize if we are to find the foods we need, eat the right amount of each, and avoid foods that are unhealthy, poisonous or rotten.

Our sensory systems interact with the hormonal and nervous pathways that detect when our blood glucose is low, or when our body is starting to burn fat. Together these mechanisms combine to regulate appetite. Every connection and every feedback loop has been shaped and reshaped by generations of selection to meet the nutritional needs of our bodies in the context of the foods that were available to our ancestors.

In pursuit of energy

Many animals struggle simply to find and eat enough food to avoid starving. Some leaves contain plenty of nutritionally available energy, but others contain largely indigestible fibrous material, especially during a drought or in winter. It is therefore not only possible, but quite common, for herbivores like the African greater kudu antelope to die of starvation with a stomach full of food in a landscape seemingly replete with plant life. They are driven by desperation to fill up with the only foods available, yet these foods do not yield enough energy to fuel the basic needs of the animal's organs.

For many animals, then, the rules of foraging are simple: ingest as much energy as possible. Biologists often think of energy as a currency, which, like money, must be acquired, stored and spent judiciously. Energy is a convenient currency both for most forms of dietary intake and for most forms of expenditure because we burn chemical energy to keep our brains functioning, hearts pumping, guts digesting, livers detoxifying, nerves firing and muscles contracting. Thus the costs of staying alive and of finding and eating food, avoiding predators and finding mates can, at least in principle, be measured in terms of energy.

Because weight gain is a consequence of ingesting more energy than we burn off or excrete, it makes good sense that our modern problems of excessive weight gain may be due to strong evolved urges to maximize energy intake. Indeed, many evolutionary accounts of the current obesity crisis only go as far as the fact that humans have historically evolved to overeat energy-rich food like honey or fatty meat—foods that were rare until last century.

At first glance it seems obvious that humans must be energy maximizers and that our major nutritional challenge today is how to overcome our evolutionary past and limit our energy intake in a world of overabundant and cheap energy. Research on foraging in other primates appears to support the idea that humans have been shaped by natural selection to maximize energy intake. For some time, data on wild primates seemed to suggest that the primary objective of the foraging monkey or ape is to meet their daily energy requirements. However, recent developments in nutritional ecology are challenging the interpretation of why primates, including humans, eat so much energy-rich food when it is available, and turning the spotlight on our need for protein.

Protein and the Peruvian spider monkey

The Peruvian spider monkey is a handsome little animal, almost entirely black, with the improbably long arms typical of all spider monkeys. Anyone who has spent time in the rainforests of Peru, Bolivia or western Brazil will have seen them exuberantly climbing and swinging in the forest canopy, or eating young leaves, seeds, flowers and especially the ripe fruits of rainforest trees. Small-bodied mammals like spider monkeys live fast-moving lives and burn a lot of energy. They also tend to gorge on ripe, sugary and fatty fruits when these rare treats become available, eating many times the total energy they would eat on a normal day when such fruits are not available. Just like humans, Peruvian spider monkeys appear to be typical energy maximizers.

Annika and Adam Felton, doctoral students at the Australian National University in Canberra, spent nine months following 15 adult Peruvian spider monkeys in Bolivia, recording what each animal ate. The monkeys ate 84 different types of food over the study, and when the Feltons got back to Canberra they analyzed the nutrient content of each of these foods. Their records of daily feeding activity revealed a surprising pattern. Each day, monkeys ate a combination of the available foods that contained very close to 45 kilocalories of protein. However the carbohydrates and lipids that the animals ate on a given day varied from 167 to 1476 kilocalories—almost a nine-fold difference. It seems that these monkeys feed first to meet their protein needs. When energy-rich but protein-poor fruits are available they keep eating until they meet their protein needs, massively overdosing on energy.

The behavior of spider monkeys makes sense given their diet. Excess energy can be stored as fat and an energy shortfall can be met by breaking down and using body fat or, as a last resort, muscle protein. While protein can be turned into useable energy, the reverse cannot happen. Most of the protein that bodies need for growth, body repair and reproduction must come from the diet and it cannot be stored as effectively as energy can be stored. So meeting the body's protein needs must come before maximizing energy—but when a sweet fatty food source like ripe figs becomes available, monkeys can both max out on energy and meet their protein needs. Spider monkeys are not energy maximizers after all.

Annika and Adam Felton's discoveries about a small fruit-loving monkey made a lot of sense to nutritional ecologists Stephen Simpson and David Raubenheimer, who worked with them to interpret their surprising results. Simpson and Raubenheimer have found that each animal species they have studied, from flies and locusts to mice, rats and humans, has a "rule of compromise" that individuals use to decide how much to eat when their diet does not have the balance that they need. Some animals, like spider monkeys, are driven to meet their protein needs, occasionally over- or under-eating carbohydrates and fats if their diet has too much or too little of these nutrients. Other animals eat primarily to meet their energy needs and their energy intake does not change much even if their food is particularly rich or poor in protein. But the most exciting thing about the free-living spider monkey data was the similarity to a study Simpson and Raubenheimer had recently published on humans.

Simpson and Raubenheimer conscripted ten healthy student volunteers for what might be the cushiest gig in university life: a six-day fully catered trip to the Swiss Alps. All that the volunteers were required to do was eat breakfast, lunch, an afternoon snack and dinner. The rest of the day was theirs to spend as they wished. At each meal volunteers could choose as much as they liked from a buffet of 5 to 15 items, each of known protein, fat and carbohydrate content. As volunteers selected each food from the buffet, their plates were weighed, and the amount of food they left on their plates was weighed at the end of the meal. For the first two days, the range of foods at each buffet meal varied. There were high-protein items like tinned tuna, baked white fish and ham, and low-protein but high-carbohydrate items like bread, honey and cous cous. Half the subjects then spent days three and four eating from buffets that were high in protein but low in carbohydrate energy: ham, cheese, eggs, tuna, cottage cheese, salmon and pork. The other half could choose only among high-carbohydrate options: food such as bread, jam, pasta, tarts and baked potatoes. After two days on these experimental diets, subjects were again given two days on the same broad menus as on the first two days.

Those subjects who spent the middle two days on the high-protein diet ate foods that were lower in total energy than they did at the start or the end of the diet when they had a broader choice of foods. In fact they ate almost exactly the same amount of protein per day as they did under free choice, but their total energy intake was lower because they ate less fat and carbohydrates. The other half of the subjects who spent the middle two days on a high-carbohydrate diet overate carbohydrates dramatically, but kept their total protein intake similar to their protein intake under free choice on the first two days. As a result the total energy content of the foods they ate was a massive 45 percent higher than it had been on the first two days. On the last two days, with free choice again, these subjects once again ate approximately 12 to 15 percent protein and roughly the same amount of total energy as they had on the first two days.

These results made sense in light of the fact that eating protein is far more likely to lead to satiation—the feeling of being full, which stops us from eating—than either fats or carbohydrates are. Perhaps it is the extraordinary variety of meats on offer at Christmas dinner that leads so rapidly to the feeling that we need never eat again. Comparisons of diets of people in different parts of the world suggest that the amount of protein eaten varies far less than either fat or carbs.

All of these strands of evidence led Simpson and Raubenheimer to propose the "protein leverage hypothesis" as an important factor in the current obesity crisis. According to their hypothesis, humans need at least 15 percent of dietary energy to come from protein in order to meet our protein needs and not exceed the energy that we use up. When the diet contains more than 15 percent protein, we eat less carbohydrates and fat and our total energy intake is lower. In fact for every excess calorie of protein we eat, we eat 11 fewer calories from carbs and fats. But the serious problems begin when our diet has less than 15 percent protein. For every calorie of protein that is missing from our diet, we have to eat a whopping 53 extra calories of carbohydrates and fats before our bodies are satisfied that we have eaten enough. This idea, although simple, is surprisingly counter-intuitive: small changes to the relatively modest amount of protein that we need to eat to satisfy our evolved hunger-regulating systems may cause the massive excess fat and carbohydrate consumption that gets so many of us into trouble.

The price of eating well

It is common for politicians, especially those most zealous about free markets, to claim that nobody is forced to eat badly. Give people a choice, they argue, and markets will find the most cost-efficient ways of delivering what people want to eat. Although people in industrialized societies certainly do have a wider choice of foods available to them than ever before, that does not mean people are getting the food they need. For one thing, our ancient evolved tastes often fool us into wanting foods that are crammed full of sugar, starch, fat and salt. For another, the economics of producing food can bias us toward buying the wrong things.

Analyses of the price of groceries by Adam Drewnowski and his colleagues reveal a link between food pricing and obesity. Energy-dense foods (food with many calories of energy per gram of weight) also cost less per calorie of energy. Healthier diets rich in highly nutritious yet low energy-density foods such as fruit, vegetables and lean meats not only cost more, but their cost has risen more rapidly over the last 60 years than the cost of energy-dense foods like cereal grains, sugar, oils and processed or manufactured foods high in starch, sugar and saturated fats.

When I came across Drewnowski's papers, I was excited: surely the common ground between environmental and genetic causes could be the quantitative science of economics. At the time I was particularly intrigued by Simpson and Raubenheimer's protein leverage hypothesis, and so I did a quick assessment of the costs of 111 common foods in my local supermarket and fast-food outlets. I was amazed that every megajoule (1000 kilojoules or 239 kilocalories) of energy from protein adds $3.26 to the average price of a food, but every megajoule of carbohydrate actually reduces the cost of food by 38 cents. Forget cheap carbohydrates, supermarkets are giving them away! Staple carbohydrates like bread, pasta and maize meal as well as sugary or starchy processed foods and sweet drinks from fruit juice to cola are cheap. At least they are cheaper than vegetables, lentils, meat and dairy. Because sugars and starches are cheaper relative to protein than at any other time in human history, economic costs are likely to bias the foods we buy and eat toward energy-rich yet protein-poor diets. Within industrialized societies this effect is likely to be most extreme for poor people who have access to a wide range of foods but who are constrained in which foods they can afford to buy.

We estimated that it would cost less than 72 cents to reduce energy intake by 380 kilocalories per person per day, the kind of reduction that would bring energy consumption back to 1970s levels. This adds up to an annual cost of $262 per obese person, or less than one-fifth the estimated additional medical costs that are spent on each obese person in America. The US Centers for Disease Control estimate that medical spending on each obese individual in the United States is $1429 higher than the spending on each person of normal weight, and half of this expenditure is borne by the taxpayer. Switching to healthier, more protein-rich foods appears to be a highly cost-effective intervention in terms of the medical costs of obesity alone.

It is one thing to show that for a relatively modest cost people can switch from cheaper foods high in carbohydrates to more expensive foods containing more protein. It is another to make this happen. One possibility is to subsidize high-protein foods such as lentils, lean meat and fish. Another is to reduce subsidies or tariff protection on sugar and cereal staples. An alternative to the politically perilous business of intervening in commodities markets is to tax products that clearly generate a large part of the public health burden. Reductions in carbohydrate intake might more effectively be achieved by raising the price of carbohydrate energy than by lowering the price of protein. Products like soft drinks, fried potato products and ice cream constitute a large proportion of the energy intake of adults and children at risk of obesity but contain little or no protein. Special taxes on cheap carbohydrates could well prove to be particularly effective.

Evidence already indicates that consumption of soft drinks, energy drinks and even fruit juices contributes in a big way to the obesity crisis. Consumer advocacy groups in the United States, Australia and elsewhere have for several years called for high-sugar drinks to be taxed, and two American states have already enacted substantial taxes of this kind. Soft-drink consumption drops quickly when prices rise, a phenomenon economists call demand elasticity. According to economic theory, demand for products that are necessities is less elastic, whereas products that consumers do not strictly need tend to show a drop in demand when prices rise. Sugary drinks are definitely not necessities, and the dramatic elasticity of soft-drink consumption suggests that taxation of high-sugar drinks and other sources of cheap carbohydrate energy may help reduce energy intake where that help is most needed. Adam Smith seems to have foreseen this in 1776 when he published *The Wealth of Nations*, a book that is to economics every bit as much a cornerstone text as Darwin's *On the Origin of Species* is to evolutionary biology:

> Sugar, rum, and tobacco are commodities which are nowhere necessaries of life, which are become objects of almost universal consumption, and which are therefore extremely proper subjects of taxation.

Natural selection throughout our history has ensured we need certain nutrients, particularly protein, more than we need others. It is early yet, but I predict that the extremely welcome convergence between evolution and economics will show that our evolved needs will cause patterns like demand elasticity. Most modern societies already recognize that the healthcare and social costs of tobacco and alcohol make them "extremely proper subjects of taxation." Might we see sugary drinks, sugar itself and saturated fats go the same way?

Do you want fries with that?

When we had our first child I promised my partner that I would never take our kids to McDonald's. It is now eight years and counting, and

I have only broken my promise once; however, I am not counting that occasion because we went to use the pleasantly hygienic amenities under urgent circumstances and we did not stay for the food. But I have a dirty little secret. Every few months, I get a mysterious hankering to roll up to the drive-through and order a burger, fries and a cola. I inevitably regret it, but until recently I have not spent much time thinking about the food.

Armed with the fact that modern humans in Western societies are estimated to need at least 15 percent of their kilocalories from protein before they fall foul of protein leverage, I went on a field trip to see how a modern junk-food diet would stack up. First, I assume that an active man of my height would need 2380 kilocalories of energy per day. This is generous, given that McDonald's bases its estimates on kilocalories per day for an average adult. The good people at McDonald's provide nutrient content of their foods on their Web site, and it may surprise you to learn that protein provides 18 percent of the energy content of the world's favorite burger, the Big Mac. Because each Big Mac contains 540 kilcalories, this means that a man like me could surpass my target for protein energy intake, without exceeding my total energy needs, simply by eating four Big Macs every day!

But the bad news for those of us who love fast food is what happens if you eat the Extra Value Meal containing a Big Mac. With medium fries and Coke, you would only need 2.1 meals a day to meet your energy target, but would still need more than three meals a day to meet your protein target. If you ate three Extra Value Meals, as your protein-leveraged appetite would have you do, you would be consuming 42 percent more energy than you need. Every day. On top of that, and returning to the price of food for a moment, even though fries and Coke in an Extra Value meal are thought to be among the most profitable items for the franchise, at my local McDonald's they reduce the cost per calorie by 23 percent compared with the Big Mac on its own. So if you love junk food in a non-negotiable kind of way, try to stay away from the fries and soft drinks—despite the irresistible value for money that they appear to offer.

I have a more important point to make than showing how to meet your protein target in a fast-food restaurant. If you include items like French fries, hash browns, and potato crisps in your diet, their low protein (4 to 5 percent of energy content) relative to their high energy density gives you a very high chance of exceeding your energy intake needs unless you go to great lengths to offset these items with foods like very lean meat and lentils. This is especially true of soft drinks; a typical non-diet soft drink contains about 7 percent of your dietary energy needs for the day, and all of this is in the form of sugars that are immediately available. So-called energy drinks are no better. Energy drinks may be useful to people playing sport or running marathons, but in the current obesity crisis driven by overconsumption of energy, their widespread availability and marketing is perverse. Unfortunately fruit juices are not all that much better. They deliver all the sugar from fruit without the fibre, and plenty of simple sugars.

Many animals that maximize their net rate of energy intake do so to minimize the time it takes to meet their needs rather than to maximize their daily energy intake. I rather suspect that humans are like this in some important respects. Our hunter-gatherer ancestors mostly sought to meet their short-term protein and energy needs and then spend their time doing more interesting and less demanding things like making tools and weapons, fixing shelters, painting on rock walls and hanging out. The transition to farming robbed many of our ancestors of their time for leisure and creative pursuits, and modern work is even worse. It is often said that modern working families run on daycare and fast food, and as a father of two young kids I must agree. I have not considered in this chapter the time costs involved in meeting our dietary needs, but time is every bit as valuable a currency as money and is just as deserving of attention. After all it is the convenience of fast foods that makes them so attractive to many of us.

The meaning of life

In ancient Greek theater, the gluttonous *obesus* was a figure of ridicule and fun. You might think that modern society is more enlightened, but obese characters in modern movies and books are often expendable or the objects of cruel humor. Think of Augustus Gloop in *Charlie and the Chocolate Factory*, Monty Python's Mr. Creosote, and the cryptically named Fat Bastard in *The Spy Who Shagged Me*. Television news stories regarding obesity inevitably screen voyeuristic footage of obese people going about their daily business while drinking super-sized soft drinks or scoffing French fries. These stories stir disgust, or outrage against the apparently gluttonous and slothful. Contempt for the obese is both rife and still tacitly acceptable in many societies. Yet are they to blame or are they victims? This is a question that courts increasingly have to weigh, as they preside over lawsuits by overweight plaintiffs against fast-food outlets.

We tend to frame questions about why people get obese in terms of a lack of willpower to exercise and to eat well, or a powerlessness to overcome the tyranny of their biology. This reflects our lazy habit of polarizing environment versus biology and culture versus evolution. If we are to truly understand the global problem of excessive weight, and if we are to do anything worthwhile to mitigate this problem, then we need to acknowledge that biology and environment are not neatly separable. It is biological bodies that get fat in response to chemical laws in the context of the food that is available and affordable in the environment. Changes to the environment can make some people's bodies become much fatter or much thinner, and yet have no such effect on other people's bodies.

In order to illustrate the complex and subtle interplay of genes and environment in obesity, we need to revisit the Pacific islanders. Their ancestors gradually colonized the Pacific over the last 8000 years, from which time until very recently they ate a diet rich in fresh reef fish and tuna. This is about as healthy a dietary foundation as you can get: high in protein and the healthiest fats. They also ate vegetables and fruit high in fiber and complex carbohydrates such as breadfruit, taro, yams, cassava and bananas; and coconuts, which are rich in the healthiest fats. Agriculture on Pacific islands was confined largely to these fruits and root crops. There was probably seldom great overabundance of food, and probably periodic hunger, including the very real chance of going days without food if a fishing trip fell foul of weather or currents. As a result, Pacific islanders are probably among the peoples whose ancestors never experienced the very high carbohydrate diets that characterize transitions to agriculture. Pacific islanders may also have adapted to the presence of abundant fresh fish, evolving a higher protein intake target than people in many other parts of the world.

Recent modernization in the Pacific reduced both physical activity and reliance on traditional foods. At the same time imports of salted and processed foods full of saturated fats and cheap carbohydrates rose. Since the 1950s, life in Micronesia has changed substantially with the rise of a cash economy fueled first by US subsidies and then by the sale of tuna fishing rights to Japan. As a result, the local foods have been steadily supplanted by rice, wheat flour, sugar, tinned fish, fatty tinned meats, and turkey tails.

The tale of the turkey tail is typical of the tragedy. Turkey tails are gristly, fatty skin flaps cut from American Thanksgiving and Christmas turkeys and either discarded, used for pet food or exported frozen to poor places like Micronesia where they are sold to the poorest citizens. The US Department of Agriculture's supplementary feeding program, which provides school lunches comprising tinned fish, tinned meats and rice, has also been blamed for increasing food dependency and replacing the production and consumption of healthy local foods with less healthy imported alternatives.

I don't have much of a palate for conspiracy theories, and goodness knows there are enough of them surrounding the modern diet. I think that obesity on Kosrae and elsewhere in the Pacific is far more interesting than a conspiracy theory could ever be. It seems likely that a very unfortunate confluence of interlinked political and commercial interests as well as global economic changes and the challenges faced by most developing nations have had the largely unintended effect of changing diets in places like Kosrae, Nauru, the Cook Islands and Samoa. The shift from a very healthy locally sourced diet rich in fish protein, healthy fats and complex carbohydrates to a diet full of low-quality tinned protein, cereal grain staples and sugar has been hastened by poverty, the transition to a cash economy and the growing influence of foreign developed nations.

This part of the story is not unique to the Pacific: wherever people make the nutritional transition from a traditional diet to a Western diet, they are suddenly ravaged by the unholy trinity of obesity, increased type-2 diabetes and metabolic syndrome. But when these people's ancestors somehow avoided the agricultural revolution, as was the case for most Pacific islanders, this transition is so much worse. Pacific islanders, Australian Aborigines, indigenous Americans and other people whose ancestors lived off few or no domesticated carbohydrates tend to be particularly at risk of type-2 diabetes and metabolic syndrome. These afflictions are caused by an inability to handle the big spikes in blood sugar that come with eating large amounts of starch and sugar. People in these populations have experienced both the transition to agriculture and the escape from hunger simultaneously, and their genes have no instruction manual for how to handle either.

If ever there were an example to counter the popular yet mostly baseless idea that culture and technology buffer modern humans from the cruelty of natural selection, it is visible in the Pacific islands, Aboriginal Australia and American reservations. The lack of historic exposure to cereal grains and sugars may make people from these areas especially prone to turning cheap carbohydrates and lipids into fat, and developing insulin resistance. The plight of Pacific islanders is probably worse than it is for many other peoples because with such abundant fish, a traditional history of high-protein intake may make the protein-leverage effect particularly strong for them. Evolutionary biology can help us to understand the human condition, but in order to resolve how we must act, we need to combine this knowledge with all the other conceptual and philosophical tools at our disposal. Global food production, trade, marketing, distribution and even our understanding of nutrition are intimidatingly complex. How we should find our way as individual foragers in a world of crippling choice and deafening messages about food is both beyond the scope of this book and beyond my expertise. But I can leave you with a few vague insights to ease the way.

First, follow the ancient Greek aphorism and know thyself. Knowing who we are and how we have been shaped by evolution can help us understand not only what we should eat but why. In the dawning era of personal genomics, the capacity to know ourselves and our ancestors is greater than ever before, and the prospect of diets individually tailored to the adapted genes of particular people cannot be far away. This is not the genetic determinism so feared by social constructionists: it shows just how useful an understanding of evolved genetic diversity can be in improving lives. Especially the lives of the poor and disenfranchised—the people most at risk of obesity and its accompanying ills.

Second, when commercial interests are involved, be they Japanese fishing fleets, fast-food chains or soft-drink companies, follow the Roman habit of asking, *cui bono?* ("to whose benefit?"). Commercial and biological forces often intersect. An informed society is one whose citizens ask and can resolve who benefits from a particular situation, law or deal, and this applies to food as much as to any other area of society.

But if I were to give you only one message from this chapter to snuggle up to it would be Michael Pollan's advice with which he begins and ends his excellent book *In Defense of Food*: "Eat food. Not too much. Mostly plants." I would probably dilute Pollan's pithiness by adding, "Make sure you get enough protein."

3

Weapons of massive consumption

Started this chapter: 1 August 2009

World population: 6,774,705,647

Revised this chapter: 2 September 2010

World population: 6,865,942,377

Human population growth and escalating consumption are damaging the capacity of the earth to sustain society in almost every way imaginable. The roots of this problem can be found in the way natural selection works by favoring those individuals who consume most relentlessly and reproduce most prolifically. Can evolution teach us anything about self-restraint?

THERE IS A SPECIAL KIND of cold that comes with waking up before it is light and heading out into the wilderness. In Botswana's Okavango Delta in July, winter winds off the Kalahari Desert sharpen the chill. I felt like I wasn't too badly off, huddled low in the wooden makoro canoe in my embarrassingly new goose down jacket. But my guide, Motsamai, was much less comfortable, poling our makoro through the reeds toward Baobab Island. No more than 300 yards at its widest, this low strip of dry in the Okavango wetlands is named for the tree that sits at its center. Baobabs are, by far, the largest succulent plants on earth.* I have seen many bigger baobabs than the 10-foot wide, 400-year-old specimen I visited in the Okavango, but as Motsamai and I warmed ourselves in the early sunlight I was struck for the first time by just how much this tree would hate elephants, if it only could.

Motsamai told me that the next closest baobab was a four-hour makoro trip away—roughly 5 miles. Baobab seeds germinate much closer than that every year, but baobabs, like many long-lived tree species, grow at a pace that would make glaciers impatient. They are also delicious to herbivores—especially elephants. Even a sturdy young baobab—perhaps half a century old—could be ripped clean out of the ground by an elephant in search of a snack. It takes decades of incredible luck for a young baobab to grow to an elephant-proof girth and height. Even then it would have every reason to fear elephants. The tree I visited had recently been gouged along the trunk from near the ground to 13 feet high by an elephant in search of the baobab's juicy inner tissues. Evidently this happens often: I counted scars from at least a dozen similar episodes over the years, in various stages of healing. Clearly the world's largest land mammal profoundly impacts the largest succulent plant.

Ecologists talk about elephants as ecosystem engineers; animals that reshape the vegetation and even the physical structure of the places where they live. That same morning, I watched a big bull elephant relentlessly shake a tall ilala palm tree to dislodge any ripe fruits, and I saw where elephants had recently dug up and eaten the roots of a sweet-thorn acacia tree, leaving it with no chance of survival. They think nothing of pushing over trees that took decades to grow. The

* The largest, a 6000-year-old specimen in South Africa, measures 154 feet in circumference. Archaeologists have found evidence that its hollow interior has, in the past, sheltered San bushmen and Afrikaaner voortrekkers. That same hollow now hosts, in inimitable South African style, a 50-seat pub.

actions of one or a few elephants may seem trivial, but as protected populations of African elephants grow, their cumulative destruction transforms woodland to grassland, dramatically degrading habitat for plants and animals alike.

Even before international traffic in ivory was totally banned in 1990, Botswana had ballooning elephant populations. Vast areas of Botswana are now paying for the success of that country's elephant conservation programs, with some areas completely devoid of large trees. These landscapes are haunted by the aging skeletons of trees that germinated and grew to an elephant-proof size in previous centuries, when elephant numbers were lower. Herds used to move over hundreds of square miles in search of the best available food, returning to some places annually, but sometimes not returning to other places for decades. An individual or a herd that destroyed the trees in a particular area would hardly have felt the consequences because they would have moved elsewhere.

Nowadays, most African elephant populations live in conservation areas and reserves, surrounded by fences or by farmland. In northern Botswana, a series of fences have been erected to prevent the transmission of wildlife diseases to livestock. More than 150,000 elephants are trapped between these fences, subsistence farms and the heavily landmined Angolan border to the north. Conservation efforts focus on opening up ancient migratory corridors, including removing landmines, giving the elephants somewhere to go until those new spaces reach capacity.

Hemming animals as destructive as African elephants into reserves can spell disaster. In South Africa's Kruger National Park, the elephant birth rate is around 5 percent per year. Populations grow like money in an account that attracts compound interest; just as previous interest payments start to attract interest themselves, each successive generation of elephants that is born also matures and starts to reproduce. The result: ever accelerating growth in the elephant population. At the time Kruger National Park was proclaimed in 1898, elephants had been hunted to very low levels. Numbers grew from perhaps a few hundred to about 6000 in the early 1960s. Between 1967 and 1994, rangers culled some elephants annually to maintain a stable population size near 7000. In 1994, however, elephant culling ceased due to pressure from animal rights organizations, and elephant numbers are now comfortably north of 12,000 and rising.

Eventually, elephant numbers will exceed the capacity of the land to support any more elephants. When this happens, starvation will prevent most animals from breeding and kill many others, and numbers will stabilize. But some experts argue that there are already so many individuals that they are permanently damaging the vegetation of the park, the soil and the community of animals that depend on both. Experts differ on just how many elephants Kruger can sustain before they cause lasting damage to one of the world's great ecological treasures, and before all Kruger elephants find themselves one drought away from catastrophic starvation. For now, Norman Owen-Smith and other world-leading elephant ecologists suggest that there is no compelling need to cull elephants on a large scale, although culling might be necessary on a small scale in places where elephants are causing too much damage.

Nobody likes the idea of intelligent, majestic and highly social elephants being culled, and fortunately for Kruger elephants, there may be a brief respite. The fences between Kruger in South Africa and Limpopo National Park in Mozambique have been removed, and these areas will soon be joined to two national parks in southern Zimbabwe, making a total area of 24.7 million acres. Decades of poaching in Zimbabwe and civil war in Mozambique mean that elephant densities there are much lower than in long-protected Kruger. As elephants move out of Kruger and into less densely populated areas, we may see some reduction in elephant densities and elephant-induced damage. But the question of culling will raise its head again.

Natural selection bites the invisible hand

Many people—including the makers of many nature documentaries—are under the impression that most animals are in some kind of harmonious balance with their environments and that the capacities for overpopulation and for destruction of one's own habitat are uniquely human failings. This view goes hand in hand with the view that prehistoric people and modern hunter-gatherer tribes possess some higher form of ecological intelligence and that environmental destruction afflicts only the post-industrial world. Even though modern humans have raised environmental destruction to a high art, we are by no means alone. The origins of our destructiveness reach far back into evolutionary history and we share with other animals and with ancient societies the capacity to outstrip our resources and to render uninhabitable the places where we live.

In this chapter I explore the origins of the human tendency to breed and consume until we outstrip the capacity of the land to sustain us. But you need only to spend time around African elephants to see how over-abundant wild organisms can devastate their natural habitat. Elephants are not unique in this regard either. Flying foxes (giant Australian fruit bats) can ravage the trees in which they roost. You can see this in action a few miles down the road from where I live. At dusk in downtown Sydney, 22,000 grey-headed flying foxes head out from their roosts in the Royal Botanic Gardens like squadrons of World War II airplanes heading out for an evening of bombing. By day, great colonies roost in just a handful of heritage-listed trees, which rapidly end up looking as though they have been through a cyclone followed by a locust plague.

The key to understanding both population growth and overconsumption can be found in the process that made elephants, flying foxes and human beings what they are: natural selection. That humans and other animals can outstrip their resources and render their home uninhabitable seems at first to be a paradox. Surely evolution is about achieving a perfect and harmonious fit between organism and environment? Surely consumption and destruction on a scale that threatens the habitat on which life depends would be eliminated by natural selection? Unfortunately, natural selection is not like design or any other human creative process that involves foresight. In reality it is a tremendously wasteful and inefficient process that is often most effective in hardship, starvation and misery.

One of the most important influences that led to the idea of natural selection was the work of the Scottish economist and demographer Thomas Robert Malthus (1766–1834), who wrote in 1798 that "the power of population is indefinitely greater than the power in the earth to produce subsistence for man." Population growth is powerful, as Malthus recognized, because members of each new generation become reproductive themselves. Thus populations will grow exponentially and rapidly outstrip the food supply, resulting in distress, starvation, disease and conflict. Both Darwin and Alfred Russel Wallace—who independently discovered natural selection at the same time as Darwin—recognized that in animal populations this Malthusian misery would strongly favor the reproduction of some individuals over others.

When a population grows but the total amount of food available to that population stays the same, there is inevitably less food available to each individual. Biologists call this competition. Competition reduces the average amount of food available to everyone, but it also results in the emergence of clear winners, who manage to find and eat more food than the losers around them. These winners live longer and produce more offspring than their neighbors. When differences in competitiveness among individuals are due to genetic differences then competition can lead directly to evolutionary change.

Every elephant alive today is the descendant of a long line of elephants that were good at finding and eating the food they needed, turning that food into baby elephants, and looking after and teaching those babies to do the same. Even when elephant numbers were high and all the individuals were hungry, those elephants best able to find, hoard, eat or otherwise monopolize the most resources left the most descendants. Those elephants that were less competitive or less fortunate in the scramble for food left no descendants. That is why today an elephant that rips up a baobab sapling has no concern for the impact that her actions have on other elephants. By destroying the tree she wins twice: once by eating whatever she can and sharing it with her genetic relatives in her herd, and a second time by denying that food to the other elephants in neighboring herds. Her actions have made her more likely than her competitors to succeed in the main business of life—turning the resources she consumes into yet more elephants, each of which carries half her genes.

Natural selection is not about making the best kind of elephant, baobab tree, flying fox or human being. It often seems that way because evolution can fashion the most exquisite match between an organism and its environment. But those of us who study evolution closely, see that it is instead a perpetual war fought on many levels and many fronts. The winners and losers in this war are the individuals themselves, but very often the process can be hideously costly for the population or the species. This is why few things make biologists more likely to have a seizure than hearing talk about the "good of the species" or "perpetuating the species."

The evolutionary interests of individuals often conflict directly with the ecological interests of the species. If elephants were interested in the long-term prospects for elephant-kind, and in one another's mutual well-being, then each would take only the food it needed. They would use their gargantuan brains and much-celebrated memories to devise schedules to harvest each baobab sustainably, optimizing the number of sturdy baobabs and other trees available. When they sacrificed an acacia, they would use their deep infrasound rumbles to call all other elephants for miles around to share the feast. And they would produce only enough baby elephants to replace those that died.

The situation sounds idyllic yet improbable. To see how ridiculous it is, imagine in this population of happy hippy elephants an elephant that decided to cheat: to eat what it liked when it wanted, to push over trees just to get a few choice branches, to keep any juicy finds to itself (while cashing in on the generosity of the more sharing elephants), and to produce as many calves as it could afford to care for. Such an elephant would enjoy a massive fitness advantage over the others. And any gene that disposed elephants to cheat would be inherited by far more calves than alternative forms of the gene. Freeloading elephants would not only increase steadily, replacing the happy hippy elephants, but like freeloaders everywhere they would be ruining things for everybody. As a result of their rampant population growth and excessive, wasteful consumption, there would be less food per elephant, and the environment could get trashed, making it uninhabitable for elephants and a great many other animals besides. That is to say that they would behave much like elephants in protected populations do today.

Massive consumption

Humans today have a lot in common with elephants in northern Botswana or South Africa. There are so many of us that our consumption is damaging the planetary systems that sustain us. We have expanded into just about every habitable part of the world, and there really is nowhere else to go, pipe dreams about Martian colonies aside. We consume the finite resources on earth so greedily that we will soon run out of many of the most crucial. Global oil production is expected by many experts to reach its maximum (peak oil) soon, and if supply slows dramatically, the rise in price will make previous fuel crises look trivial by comparison. There is great interest in how we will wean ourselves from our dependence on oil for energy and for the production of incredibly useful petrochemical products like plastics; something we will probably need to do long before the last barrel of oil is ever extracted.

And global stocks of just about every edible fish are so damaged that even the more optimistic fisheries scientists doubt that many species can recover. The overconsumption of resources such as oil and cod drives consumers to switch to other resources where substitutes can be found, and in turn to overexploit them. Some resources, like oil, are strictly finite. Reaching peak oil necessitates a switch to other energy sources, but coal, natural gas and current nuclear fuel isotopes are all ultimately finite—they will definitely run out at some point if we keep using them. Less than a century ago, North Atlantic cod were so abundant and produced so many eggs that people thought stocks were inexhaustible. Yet the collapse of cod fisheries has driven many people for whom eating cod is a way of life to switch to hake, pollock, whiting, skate or even dogfish. So much so that stocks of those fish are now threatened. Fish stocks, unlike oil, are renewable resources, but they need to be well-managed or they will be unable to withstand the pressure. The way in which we harvest and process resources and deal with the resulting waste is harming our health and damaging the capacity of ecosystems to produce those resources that are renewable.

On the bright side, an ever-growing number of people are aware of the damage we are causing, and efforts to deal with the vast and complicated problems that we are creating have a higher profile now than at any time before in history. The growing panic over the climate consequences of fossil fuel use, land use change, livestock production and many other processes that generate greenhouse gases is the highest profile expression of the problem. But global warming is only the tip of an iceberg of profound ills that include air and water pollution from every imaginable toxin, ocean acidification, land clearing, dry land salinity, overuse of fresh water, and overgrazing. The problems are real, urgent and not going to go away if we ignore them or deny that they exist.

Resource consumption by human society and the resulting damage is a product of the number of consumers and the amount they each consume. Poorer nations, especially those currently undergoing rapid economic development, have the highest population growth rates. Yet their people each consume fewer resources and thus produce less waste, including lower volumes of greenhouse gases, than people in industrialized nations. Wealthier nations, on the other hand, have much lower birth rates and their populations are growing slowly or not at all, yet it is in these nations that excessive consumption is most widespread.

Consider the two largest producers of greenhouse gases in the world: America and China. The United States has less than a quarter of the population of China, yet produces just over four times as much greenhouse gas per person per year. As a result, they are very closely matched in total greenhouse gas production at slightly over 7.7 billion tons per year. The Chinese, understandably, aspire to the lifestyles that they could achieve with an economy like America's. The phenomenal growth of the Chinese economy fueled a 120 percent increase in Chinese greenhouse gas emissions in a decade. As populous nations like China, India and Brazil develop, we can expect the global output of carbon dioxide, as well as just about every other pollutant, to go through the roof.

The relationship between economic development of nations and the two aspects of their overall ecological footprint, population growth and per capita consumption, remains the single biggest obstacle to any global solution. Some people, particularly in richer nations, tend to believe the key to dealing with our environmental problems is for people in developing nations to have fewer babies, arguing that ecological damage scales with the number of mouths that need to be fed. By contrast, many developing nations argue that the egregious consumption of the wealthiest minority, particularly people in developed nations, should be curbed and their sumptuous and decadent lifestyles should be scaled back. It should be obvious that we urgently need to do both, yet the divide between developing and developed economies routinely derails discussions on how to tackle questions of sustainability—most recently and spectacularly at the dismally unsuccessful Copenhagen climate conference in December 2009.

One of my aims in writing this book is to show how evolution and economics interact to shape important aspects of modern human life. I could not have picked a more charged topic than the question of population, and my perils begin with the polarizing figure of Malthus. His 1798 *Essay on the Principle of Population* is one of the most important publications in the history of both economics and evolutionary biology. At that time the Industrial Revolution was beginning to dramatically reshape British society. Malthus recognized that the improvements that were transforming agriculture and food production would provide only temporary relief from starvation because the population would rapidly outgrow these gains. He predicted a miserable future, beset by starvation, famine, disease and conflict. The extent to which Malthus was right has been fiercely debated ever since, and much disagreement remains. Human history is littered with famines, plagues and wars that were caused at least in part by local overpopulation. Yet the industrialization of farming in Britain and Europe outstripped population growth, and trade brought in food from less densely populated countries.

Often, when human population growth has threatened Malthusian misery, improvements in food production have provided relief. After World War II, high birth rates combined with improved hygiene and medical care in many countries like Mexico and India created the fastest worldwide population growth in human history. These countries grew so rapidly that they raced to the brink of catastrophic starvation. But thanks to the so-called "Green Revolution," massively improved strains of crops yielded enormous gains in food supply, feeding these nations and allowing them to grow in both population and prosperity. Many economists today paint neo-Malthusians as pessimistic crackpots, unable to comprehend the breadth of human genius or the power of markets to deliver innovation where it is needed most. The same argument underpins the skepticism in some economic circles about "peak oil." As oil becomes more valuable and technology continues to improve, so oil reserves that were previously unprofitable and therefore unavailable become viable, often also improving supply and keeping prices from spiraling upward.

While I concede that we should never assume that technology will remain static, the tendency to pooh-pooh ideas like Malthusian limits and peak oil is also ignorant. For one thing it is an undeniable physical fact that there is a finite amount of oil in the ground. Even if the amount of oil that it is economical to extract is a moving target, if oil consumption continues to grow we will inevitably hit a peak in production. If other technologies like solar or clean hydrogen fuel make oil less valuable, or the true environmental costs of burning fossil fuels ever get built into the oil price—thus driving down demand so far that oil companies start extracting less of it—then we will have passed peak oil. The same is true of population growth. Nobody knows exactly how many people the earth can sustain, but it is absolutely clear that the number is finite for a particular level of consumption. Changes in patterns of consumption, technologies that affect the efficiency of agriculture, and the damage we are already doing to the earth make the carrying capacity for humans on earth, whatever it is, an ever-shifting one.

Most evolutionary biologists and their close colleagues, ecologists, focus on the finite nature of most resources because there are hundreds of examples where animal populations reach densities so high that plagues, conflicts and starvation set in. Population growth usually slows as a result. We need only get part of the way to the maximum possible human population before the Malthusian plagues of famine, pestilence and warfare kick in. In *Collapse*, an excellent study of why societies fail, Jared Diamond shows that overpopulation and overconsumption have repeatedly led to human tragedy of exactly the kind Malthus foreshadowed. In one of the most thought-provoking chapters he convincingly argues that population growth within a finite area of arable land contributed to the 1994 genocide in Rwanda, a conflict that is too often written off in the West as an ancient tribal conflict.

A global human population that lives within the means of our planet is not only necessary to avert environmental catastrophe, but is also essential to alleviating poverty, suffering, social injustice, and the rise of the most dangerous forms of religious fundamentalism. Nonetheless, many economists and politicians remain focused on delivering economic growth via growing populations, increased consumption and, therefore, increased production. To biologists, the idea that the earth's carrying capacity is ever growing due to increased efficiency and innovation is really just messing around with the finer details. Or as one of the greatest living biologists, E. O. Wilson, says, "The raging monster upon the land is population growth. In its presence, sustainability is but a fragile theoretical construct. To say, as many do, that the difficulties of nations are not due to people but to poor ideology or land-use management is sophistic."

Tragedy

Imagine a fleet of fishermen working on a lagoon that has plenty of good-sized fish. The fishermen only need to fish for a few hours each day to get enough to feed their families. The good news is that this much fishing does not deplete the fish—enough remain to breed and the fry have plenty of time to mature into good sized fish. The bad news is that fishermen care much more about their own welfare and status than they do about the other fishermen. It is in each fisherman's interests to stay out as long as the weather is good and catch as many fish as he can. In this way he can make enough money to steadily accumulate some wealth: a new boat, good nets, a satellite dish and television. He may be able to attract a more nubile wife and raise a larger family than he would otherwise have done. As each fisherman starts to fish for longer and to build wealth, however, something soon starts to go awfully wrong.

The problem is that the big fish are also the parents of the next generation of fish. Each big fish that is removed from the lagoon reduces the number of reproductively active fish and, therefore, the number of fry that hatch in the next reproductive season. Once fishing starts to make a dent in the breeding population, the stock can slowly dwindle or it can collapse suddenly. As a result, the fishermen have to work harder and harder to catch ever smaller numbers of ever more pathetic fish. Forget about the satellite dish, most of the fishermen and their families will starve if they stay in the fishing business. Even if the fishermen do recognize the problem of overfishing, what are they each to do about it? It does not pay any one of them to show restraint and take fewer fish if the others are going to keep fishing as much as they always have. If they are all going to benefit from restrained fishing then they need to cooperate, and ensure that nobody cheats.

The essence of the problem is identified by a very simple economic analysis. For every fish that a fisherman catches, there are two costs: the cost of spending the time and effort it takes to catch the fish and the cost of the fish no longer being part of the breeding population. While the fisherman who benefits from catching a fish pays the first cost, the second cost is shared by all the fishermen. This second kind of cost is an example of what economists call a negative externality—that is a negative consequence to somebody external to a transaction, in this case all the other fishermen. The trick, if fishing is to be sustainable, is to find a way of making the fisherman who catches a fish, and anybody who buys the fish from him later on, pay the true cost of the fish. In the language of economics, that cost must be internalized.

While my example is a fictional lagoon, this same scenario has played itself out to some extent in almost every fishery that people have ever exploited. Mark Kurlansky's exceptional book *Cod: A Biography of the Fish that Changed the World* gives a wonderfully readable history of the overexploitation and collapse of all the major cod fisheries in the North Atlantic as well as the economic, social and political circumstances of these fisheries—once thought inexhaustible—and of their eventual implosion.

The problem of negative externalities is by no means confined to fisheries. In fact it is an issue that pervades economics, psychology, sociology, political science, ethics and evolutionary biology where it contributes to a problem known as the "tragedy of the commons." In an influential 1968 essay, the ecologist Garrett Hardin likened the problem of human population growth, and the resultant growth in consumption and pollution, to the problem faced by herders grazing cattle on a pasture that is open to all (the commons). It is in the economically rational interests of each herder to keep as many cattle as possible on the land without regard to the effect this has on the land because the herder alone benefits from the sale of each animal, yet the overgrazing costs (negative externalities) of keeping each animal are shared by all users of the commons. Each herder seeks to maximize his own gain, ruining the commons for all.

One thing that makes Hardin's essay so interesting and so remarkable even after 40 years is the seamlessness between the way he conceives *evolutionary* self-interest and economic self-interest. The herders in his commons analogy, like the fishermen on my fictional lagoon or like businesses that pollute, are acting in economic self-interest. Hardin's main argument, however, was that humans, in breeding, are acting in *evolutionary* self-interest in the same way as elephants acting in their individual evolutionary interests can trash the African landscape they inhabit. Human populations explode because each parent benefits fully, in an evolutionary sense, from the birth and subsequent well-being of each child they have, but most of the environmental costs are shared by everybody.

Overconsumption, noise, and—my pet hate—excessive advertising, are rife because consumers, polluters and advertisers reap the full economic benefit of these actions yet the costs are distributed among all of us. All of these costs, be they costs of population growth or of commerce, are negative externalities. Evolutionary and economic self-interest here are no longer analogous, they are interchangeable, and for a brief moment they are one and the same thing. Just as all but the most laissez-faire libertarians recognise that the economic interests of individuals can conflict with the interests of society, so the evolutionary interests of the individual parent often conflict with what is best for human society as a whole.

Self-interest and self-restraint

Self-interest is one of the most important areas of common ground between evolution and economics, and it is also the cause of some of the biggest public relations problems that both disciplines have. Critics of both natural selection and market forces focus on the fact that both of these bottom-up processes appear to promote narrow self-interest at the expense of the interests of society, often arguing that only top-down control and regulation can effectively restore the greater good. Certainly, natural selection and market forces are both relentlessly efficient at favoring self-interest. Fortunately the critics' notions of self-interest are often so narrow as to be unhelpful caricatures; self-interest need not mean selfishness.

Among animals, and in human society, much of what appears to be altruism, kindness and cooperation is driven by underlying and often invisible self-interest. Wonderful examples of cooperation have evolved many hundreds of times over, and even simple organisms like bacteria have evolved ways of cooperating to avoid commons-like tragedies. Similarly, economists already know how to deal with negative externalities, and it is often a matter of finding appropriate and politically tenable ways of applying this knowledge. If we are to deal with the environmental challenges that arise from overpopulation and excessive consumption, then we will need to understand both evolutionary and economic self-interest and how they can be, and often are, manipulated and sometimes overcome.

Although natural selection tends to favor traits that benefit individuals, even when those same traits harm other members of the same species, this should not give oxygen to the idea that evolution inevitably favors selfishness. Earlier in this chapter I argued that natural selection tends more effectively to favor the interests of the individual over those of the group or the species, and understanding this point remains an essential step toward understanding evolution. But this is not the whole story. There are two great forces that bring the interests of individuals closer together and that often lead to cooperative and deeply unselfish behavior: relatedness and reciprocation.

In the great cooperative insect societies, like termites, ants and honeybees, workers commit what appears to be evolutionary suicide by completely foregoing reproduction to play a small and thankless role in the life of a colony. By so doing they help their close relative, the queen, produce reproductive offspring. The workers' strategy is a sound one because their own lost reproduction is outweighed by the enormous number of close relatives that go on to found new colonies as a result of their efforts. Relatedness influences human affairs, too. Kinship is the tie that binds the most important and close-knit human groups: families and small tribal groups like those of hunter-gatherer peoples. Sacrifices are made by parents, grandparents, aunts, uncles and cousins to create opportunities for young family members to thrive and establish themselves.

Animals are also much more likely to help one another out if there is a chance their generosity will be reciprocated. This often happens when individuals live in groups and encounter one another often. South American vampire bats live in large communal roosts where neighbors get plenty of chance to know one another. They tend to forage alone, however, because of their rather strange and gruesome diet. Each vampire bat tries to find and land near a large sleeping mammal or roosting bird, creeping up to it with the kind of stealth you would never expect from a walking bat. It then nicks its victim with its teeth and laps up the blood, with anticoagulants in the bat's saliva preventing the blood from clotting.

Successfully finding and feeding off a suitable donor is a high-risk business, and vampire bats that do not succeed on a given night risk starving to death before they can find a victim the next night. Often a bat walking this energetic tightrope will, once back at the roost, approach a neighbor that has successfully gorged on a big meal in the hope that the neighbor will generously regurgitate some of the blood as a charitable donation. This system works because the small amount of regurgitated blood is worth more to the starving bat than to the full one just as your spare dollar is worth more to a street-kid than it is to you, and because bats remember who has been generous enough to donate in the past. Animals that have donated are more likely to be saved by their neighbors than those who have not donated in the past, making for an elaborate, if grotesque, system of reciprocation.

Humans live in the largest and most complex societies known to science, and much of our behavior revolves around deciphering the intentions of others, deciding whom to trust, reciprocating favors, sharing information and exchanging goods. All of this activity has shaped the evolution of the most remarkable biological structure in the history of life on earth: the human brain. In shaping our brains, evolution has also molded human society, morality and the underlying ideas regarding selfishness and cooperation through which we view both natural selection and economics. Human societies are rife with conflict and cooperation based both on allegiances of relatedness and reciprocated debts and favors. Money is a way of rendering complex debts in material form so that we can trade not in simple barter systems with other individuals, but in more complex ways to get exactly what we want in markets of up to several million individuals.

My point here is that although evolution and economics both favor behavior that maximizes self-interest, this is not the same thing as selfishness. Relatedness, reciprocity and reputation are all agents for good, allowing individuals to cooperate and trade in mutual self-interest. They might be vampire bats trading regurgitated blood, or people buying and selling their household goods on eBay. In this way, both evolution and economics have found ways to thwart self-interest and avert tragedies of the commons.

If human beings are to avert catastrophic population growth and consumption, then we need to restrain our evolutionary self-interest not only by having fewer babies but by making sure that each individual exercises much greater restraint in how much and how wastefully we consume. In his paper "The Tragedy of the Commons," Hardin was far from optimistic that we could achieve this. For many of Hardin's intellectual heirs, things have only become worse in the intervening four decades. They argue that we cannot rely on people to make the right decisions when it comes to looking after public goods, and there is no more public nor more valuable a good than our planet. Hardin and many since him have argued that the evolutionary payoffs of reproduction and consumption are so profound, having been shaped by natural selection over millennia, that our internal reward systems will naturally oppose any measure of self-restraint we show in our reproductive behavior. That is to say people need to explicitly agree not to have too many children because we will not otherwise find ways to restrain ourselves. He called this kind of agreement a "fundamental extension in morality" because it would require us to develop from reason alone a set of guidelines for acceptable reproductive behavior that was unprecedented in human history. These guidelines would be "mutual coercion, mutually agreed upon," that is, an agreed social contract to renounce the freedom to reproduce.

Top-down regulation of reproduction may have worked to reduce family sizes in the one-child policy of Communist China, yet it is way too heavy-handed to be politically tenable in most industrialized democracies. It also has many tragic yet unforeseen consequences, one of which is the subject of chapter 8. Likewise, mutual agreement is unlikely because any group that chooses to regulate its reproduction will face the threat of being outnumbered by those groups that do not choose to self-regulate.

Hardin's essay on the tragedy of the commons is one of the most profound and important works in the history of ecology and it remains as rich and relevant today as it was in 1968. He seamlessly showed how both evolutionary and rational economic self-interest can drive us along the same destructive trajectories. We may well need "a fundamental extension in morality" in order to make decisions and take action beyond those our adapted minds and our free markets have the capacity to generate, especially given the evolutionary inertia that we inherited after thousands of years of frantic reproduction and consumption. However, like many others who have continued his line of thinking, Hardin was too pessimistic about humankind's capacity to overcome this evolutionary inertia, and even over the nature of that inertia itself. For one thing, many countries have, for some time, been producing so few babies that their populations have been stable for decades. And that is the subject of the next chapter.

4

Dwindling fertility

Success in the population field, under United Nations leadership, may, in turn, determine whether we can resolve successfully the other great questions of peace, prosperity, and individual rights that face the world.

—George H. W. Bush, 1973

In many parts of the world, birth rates have already been plummeting for decades, and in some places fewer people are being born than are dying. This drop in fertility comes about because of evolved ways in which humans respond to changing economic and environmental circumstances. There is only cause for limited optimism, though, because reduced fertility seems to be tied to increased consumption.

THE CANE TOAD MIGHT WELL be the ugliest creature that ever evolved. It is also one of the most prolific. A single female can lay a phenomenal 35,000 eggs in a night, making the cane toad a perfect and most unwelcome invader. Originally from Central and South America, cane toads have overrun many Caribbean, Philippine and Hawaiian islands, and a large part of northern and eastern Australia. In Australia they were first released in 1935 to control a beetle pest that was ravaging cane crops; they have since waged a biological blitzkrieg across all but the driest parts of Queensland and the Northern Territory. They are currently racing across the north of Australia at over 30 miles per year, wreaking hideous biological destruction. Predators like monitor lizards, snakes and even crocodiles—whose ancestors had never before seen a cane toad—encounter a large, fat, slow-moving morsel and cannot resist attacking it. Unfortunately, the toads are powerfully poisonous and the predators die a very sudden and painful death.

So how is it that slow and steady breeders like humans, elephants and whales can continue to exist in a world where they are outbred many millions of times by animals like the cane toad? This question has one obvious answer in that big animals do different things and use different resources than cane toads do, often in ways that depend on their size and their intelligence. But what stops individual humans, elephants and whales from producing as many offspring as they physically can? The answer is that with reproduction as with modern design, less can often be more. Or at least, fewer offspring can sometimes be evolutionarily better.

A cane toad mother-to-be lays tens of thousands of eggs because she is playing in an evolutionary lottery. She's up against some stiff odds: the pool where the eggs are laid may dry up or stagnate; and the eggs, tadpoles or young frogs may be physically damaged, preyed upon (even though the eggs, like the toads, are violently toxic) or fall victim to disease. She ups her chances by flooding the market with eggs because there is little else she can do to improve the chances that any one of her eggs will grow up to become an adult. You can't determine the Lotto numbers drawn, but you can buy as many tickets as possible.

For humans the game is less like a lottery. We can improve the prospects of each of our babies by feeding them well, providing for them and teaching them what we know. We choose to invest more in the quality of our offspring than any other animal that has ever lived. We are down one far end of a continuum that evolutionary biologists call the *quality–number trade-off*—and cane toads are at the other. Evolution has shaped us to invest heavily in every baby we have, to the point where we trade away the option of having many kids. So unlike cane toads and most other animals, we invest much more in the quality of each of our young than we do in quantity.

Human babies are so expensive because investing in individual babies, children and teenagers has proven to be a very effective way of making grandchildren, great-grandchildren and so on. By expensive I mean that parents and relatives put a lot of time, effort, energy, love and money* into raising each daughter or son. Large, intelligent and highly social animals like humans are often relatively helpless when they are born and it takes a long time until they can fend entirely for themselves. There is also a lot to learn. It takes elephants more than a decade to reach sexual maturity during which time they learn from their mothers and other female relatives in the herd. It takes even longer for human children to grow to sexual and social maturity. While some of the undergraduate students I teach leave home at around 18 years of age, it is not uncommon for some to stay in the family unit well into their thirties!

The trade-off between offspring quality and number gets really interesting when we consider the reproductive decisions that individual people make. Just as a shopper with $5 in his pocket has to decide between buying two loaves of bread or 8 ounces of meat, so a parent must decide how best to use the limited resources he or she has available for reproduction. When people have more babies they can't give each child as much care as they would have done if they had fewer children. Humans are remarkably responsive to changes in the costs and benefits of producing children. As it becomes more costly to raise each child, people tend to have fewer of them so that they can afford these costs. The way we respond to our changing circumstances has been shaped by natural selection, and it is here that hope for our escape from overpopulation lies.

* Money is really a convenient way of storing time, effort and energy in a way you can carry in your pocket. Whether it can buy love remains debatable.

Population size throughout history

For most of history, the human population hardly grew at all. Some small bands of people flourished, but others dwindled and petered out. As a result, humans came close to extinction many times. For more than 100,000 years we would have featured on any list of endangered species that had been around at the time. Around 70,000 years ago drought very nearly pushed our species into oblivion; genetic evidence suggests that as few as 2000 people were alive in Africa. From that time until the start of agriculture, as humans spread out from Africa and around the world, the world population grew at the glacial rate of around 0.4 percent per century. As a result, no more than 1 million people lived in the entire world when the last ice age ended around 12,000 years ago.

Why was early human population growth so slow? Ancient hunter-gatherers had birth rates somewhat higher than in industrialized societies today, but people died much younger. Hunting and gathering is not a particularly economically rewarding way of life. Gatherers work hard to collect food like tubers, roots, berries and other wild fruits and vegetables that deliver very little energy relative to the effort and time spent gathering and preparing them. Ancient humans mostly took the low-hanging fruits and berries and the shallow-buried tubers and then moved on—a style of foraging some ecologists call "skimming the biomass." Traditional fishing and hunting also yield small returns relative to the work involved. So the amount of energy available to hunting and gathering humans from day to day was usually modest, limiting the number of dependent children that a family could support at any time.

Small children also come at some quite significant cost to a family of hunter-gatherers. Breastfeeding and caring for young children reduces the time and effort a mother can spend foraging. On top of that, a mother can only carry a single infant or young child in her arms or on her back. So births must be sufficiently far apart that an older sibling is able to keep up with a foraging party by the time a younger sibling is born.

There is good evidence that breastfeeding is a natural contraceptive for exactly this reason. Women hunter-gatherers, famously including !Kung women from the Kalahari desert, continue to breastfeed for up to four years, keeping births spaced far apart and population birth rates modest.

Throughout our hunter-gatherer past, life was short. The proportion of infants that died from disease, starvation, predation and sometimes from infanticide was heartbreakingly high. On top of that, famine, disease and violence stalked everybody, ensuring that, compared to modern life, hunter-gatherer population growth was slow. But the invention of agriculture propelled faster population growth.

Over 10,000 years from the end of the last ice age until the start of the common era, the number of people worldwide swelled from around 1 million to about 500 million. This number then doubled to 1 billion around the year 1800. Agriculture led to higher birth rates because each child was probably both more beneficial and less costly to the parents than it would have been in a hunter-gatherer society. Each child grew into a labourer who could work in the fields or tend the animals. More hands meant that more land could be cultivated and more livestock kept. In southern Africa today, any trip along dusty back roads is punctuated by stops to carefully circumnavigate the cattle that graze obliviously along roadsides. The cowherds are inevitably young boys, and many rural families choose to keep their young sons out of school to guard and care for the cattle. As cattle are an important form of familial wealth in much of sub-Saharan Africa, the cowherd can be an important warden of the family fortune. And as children grow, their labor becomes more valuable and so does their contribution to family wealth.

The costs of rearing children also declined as humans settled into a more sedentary agricultural lifestyle. Every agricultural development added to the amount of food available. And living in permanent or semi-permanent settlements would have allowed families to pool their childcare, leaving many young children in the care of a single relative or friend. On top of this, the constraint of having to carry young children effectively disappeared as people settled down and tended local lands and livestock. As agriculture pushed up the benefits and pushed down some of the most important costs of having additional children, family sizes settled on a new, higher, optimum. People would have had as many children as they could afford to raise, allowing for the fact that some children die in childhood. Yet the trade-off between family size and child survival still limited the number of children each mother had.

A recent study of subsistence farmers living in northern Ghana today illustrates this trade-off. Like many of our farming ancestors, modern-day Ghanaian subsistence farmers are extremely poor, largely illiterate, and have only the most basic healthcare facilities. Fewer than half of all children have received any immunizations. Women have large families, bearing six to seven children on average. Yet each time a mother has another child the survival chances of each of her existing children goes down by 2 to 3 percent. Even though people in these communities prefer having large families, having too large a family limits the care and attention each child needs in order to survive and thrive, and this constrains families from getting even bigger.

The wealthy reproduce

Agriculture changed much more than what people ate and how they got hold of it. For one thing, it gave rise to personal wealth. Hunter-gatherer societies consisted of small bands that shared food and were remarkably egalitarian in structure. Foraged food does not keep for long, so good fortune is shared and sharing is reciprocated later on. Studies of modern foraging peoples show that there are good hunters and efficient gatherers, but everybody's contribution is necessary and there is very little scope for despots or freeloaders. But the fact that grain and some other farmed foods can be stored changed all this. The existence of distinct growing seasons and hungry seasons, meant that the development of storage methods was as important in raising food availability as the farming methods themselves. Accumulated grain stores were among the first real forms of wealth, and parents who produced more children prospered because the additional labor brought more storable food, and this wealth could be transferred in time to the offspring, allowing them to prosper and proliferate in turn.

Likewise the domestication of livestock allowed people to own the animals that provided dietary protein. Excess grain and livestock could be traded for other food, goods or services. This wealth and the systems of trade that arose led to the origin of non-farming specialist roles like toolmakers and smiths, who traded their handiwork for food. Wealth that can be stored and traded can also be stolen, and the need arose to defend both wealth and the land that produced the wealth. Fathers, sons and brothers formed coalitions to defend one another's interests. Groups of neighbors or kin who could repel raids from other groups, or who could raid neighboring groups themselves, thrived.

Those who could not organize themselves well succumbed to those that did. This required organizers and strategists, giving rise to elites who oversaw defense, kept peace within the group, and took on various administrative roles, for a tax. Priesthoods also arose, exploiting people's capacity for superstition and living as social parasites off the work of farmers. This same process of organization led gradually to the growth of towns, and then cities as well as the generation of ever greater disparity between the poorest and the wealthiest members of any given group. In short, agriculture gave birth to money, and to inequitable distribution of money, and thus to all of the evils of which money is root.

Birth rates between the start of agriculture and the Industrial Revolution were prodigious compared with pre-agriculture foraging societies and modern industrialized societies, but population growth was sporadic. Periods of rapid growth and periods of equally dramatic decline largely canceled one another out. During good times, more babies led to more workers, which meant that more area could be cultivated or grazed, which led to more food, greater wealth and yet more babies.

This cycle of ever-increasing reproduction and productivity cannot go on indefinitely. Population growth in agricultural societies can only be sustained as long as there is more land to farm. When good pastures and arable land are all used up, population growth precipitates starvation and Malthusian misery unless innovative new practices or tools can improve efficiency. Throughout history, each agricultural innovation that has improved food production drove population growth. The more successful agricultural societies became, the more rapidly their populations grew, often until they outstripped their resources.

It is important to remember that it is not societies that breed, but individual mothers and fathers. As the economic inequality that comes with agriculture increased, so did the difference in reproduction between the wealthy and the poor. The wealthiest families had enough food to reproduce whereas the poorest parents could barely survive. Poverty was a population sink. Only those people who were born wealthy and remained so, or those who were intelligent or strong enough to become wealthy, contributed consistently to population growth in any meaningful way. The wealthy families that too enthusiastically turned resources into children often became the next generation's poor.

In this "reproduction of the wealthy" scenario, the wealthy reap the benefits of their own increased reproduction and yet the costs of all those extra mouths to feed are borne disproportionately by the poor—these costs are negative externalities. When this happens, the tragedy of the commons cannot be far away. In short, poverty combined with subsistence or small-scale agriculture or other circumstances where children can be put to work is a perfect trigger for population growth. In pre-industrial agricultural societies the three horsemen of Malthusian misery—famine, disease and violence—grew ever more relentless with population density.

As agriculture spread, the change to a high-carbohydrate diet led to malnutrition (see chapter 1). Higher densities and more wealth may also have stirred violence among individuals and between groups. The higher densities of people also led to more prevalent and rapidly transmitted diseases, especially as communities prospered and grew into towns and then cities. Early towns and cities had enormous public waste problems, and large populations made effective reservoirs for infectious diseases like measles. The first plague pandemic in the fourteenth century alone killed about half the people in Europe, and recurrent outbreaks of plague kept European population size from growing appreciably for more than three centuries.

The paradox of declining fertility

In the last 200 years or so, something quite unprecedented has happened, or started to happen, in every agricultural or industrializing society. These societies experience sudden improvements in life-expectancy, followed, within decades, by a dramatic drop in birth rate. These changes are known simply as the *demographic transition*, and the consequences are earth-changing. The drop in death rates is a large part of the reason the world population has grown so explosively over the last two centuries. The drop in birth rates that typically comes a generation or two later may yet prove our salvation from the unfolding population crisis.

Improved life expectancy is easy to understand: technical innovations in agriculture, medicine and public health prevent deaths at every age. As societies make the economic transition to industrialization the resulting wealth finances these life-prolonging innovations. One-fifth of the world's people currently get by on less than $1 a day; these are the people who remain most vulnerable to hunger, disease from poor sanitation and almost zero medical care. These people are concentrated largely in sub-Saharan Africa, in countries like Burundi, Rwanda, and Cameroon, as well as some Asian countries like Laos and Middle Eastern countries like Yemen. In recent decades, food production, sanitation and health systems in these countries have started improving and death rates have begun to wane. It is from places like these that most of the world's population growth in the next few decades will come.

The years from the mid-1950s well into the 1970s saw the most explosive period of population growth in human history. Worldwide, population grew at around 2 percent per year, with the peak growth rate around 2.3 percent. Medical and sanitary improvements in countries like India and Mexico enabled much of this incredible growth. Many experts expected the population explosion in these countries to inflict Malthusian misery on a scale never before seen, but the predicted catastrophic famine was largely averted by a series of prodigious improvements in crop productivity now known as the "Green Revolution," which in turn allowed for further growth.

Since peaking at 2.3 percent per year in 1963, world population growth has plunged to a much more modest 1.1 percent annually. Fertility in the most industrialized nations, including most of Europe, Canada, the United States, Japan, Australia and New Zealand, has been dropping for longest, and in some of these countries fertility is so low that populations are getting smaller. At some point between 1800 and 1970, each of these countries reached its peak birth rate and has declined ever since. In the United States, for example, the average woman around the year 1800 had seven babies in her lifetime, which is about the same as a typical woman in modern-day northern Ghana, but this total fertility rate had halved by 1900 and now sits at around two. Women in the United States are not having enough babies to replace the people who die, and US population growth is driven by immigration.

Malcolm Potts puts a thought-provokingly human face on the demographic transition by considering Charles Darwin's family and descendants. Darwin had ten children, of whom three died in childhood. If every generation of Darwin's descendants had been similarly prolific, then one would expect him to have had 49 grandchildren, 343 great-grandchildren, 2401 great-great-grandchildren and 16,807 great-great-great-grandchildren. In 2009, to commemorate the bicentenary of Darwin's birth, London's *Daily Mail* newspaper estimated that there are around 100 living great-great-grandchildren and great-great-great-grandchildren of Charles Darwin and his wife Emma. This is a much more modest number than the thousands that we might expect if every Darwin descendant left an average of seven adult offspring. The Darwin family provides such a striking example of dwindling fertility because at the time Charles and Emma were having their children, between 1839 and 1856, fertility seriously began to decline in England.

What role does evolution play in the fertility decline? The change itself cannot be a direct response to natural selection for smaller families. For selection to alter any trait, especially something as important to fitness as offspring number, it must usually act over tens of generations at least. But the change occurred almost completely within three or four generations. Instead, evolutionary biology can help us to understand the demographic transition by illuminating how changing environmental conditions alter when people reproduce, how many babies they have and the kind of care they lavish on those babies.

Unlike the decline in death rates, the decline in fertility does present an interesting paradox for the evolutionary biologist. Why would an organism cut back on the number of babies it has under the most favorable circumstances its species has ever encountered? Why has the incredible wealth generated since the Industrial Revolution not been translated into the maximum number of babies that could possibly be born and raised? The paradox thickens when we consider that somewhere during the demographic transition, the relationship between wealth and fertility flipped; not only are people in the wealthiest countries less fertile than those in poorer developing countries, but within the wealthier countries wealthy people have fewer children than poorer people. Where wealth has always translated directly into more offspring, quite suddenly the wealthy families start producing fewer babies.

And this familial downsizing caught on—spreading within a generation or two from the wealthy all the way to the poorest people in those societies, although in most developed or developing nations, the wealthiest families still produce fewer offspring than the poor. Any explanation of the demographic transition must explain exactly how and why reproductive restraint has spread so rapidly, and why it is so strongly associated with national and individual wealth.

Before I go on, I should say a few words about cause and effect. Wealth might be associated with below-average fertility if wealthy families breed less, if families that breed more become impoverished, or both. If a zoologist wants to understand the relationship between resources and reproduction, she can feed some animals more than others. Or she can manipulate reproductive effort, for example by taking eggs from the nests of some birds and putting them into the nests of others, forcing the parent birds to raise either fewer or more chicks than they had laid eggs. Demographers and economists don't have it so easy; they can't experimentally impoverish or enrich some people and they can't change the number of children a family has to raise. As a consequence, they have to wait for appropriate data to become available to test their theories. In most of the rest of this chapter I consider what would cause financially better-off families to breed less because for some decades there has been solid evidence that improved wealth directly *causes* reduced fertility.

A very high proportion of pregnancies in developing countries are unintended and unwanted, at least on the part of the mother, and high fertility through lack of access to family planning, birth control and abortion can make an escape from poverty much more difficult and less likely. So it seems that fertility and wealth are interlinked, and that high fertility and constrained access to ways of managing fertility can lead to a vicious cycle of deepening poverty and rising fertility and thus to Malthusian misery. On a more optimistic note, however, reduced fertility—such as aid programs that provide access to contraception and abortion—can help families escape poverty and this escape further reduces fertility. Naturally, demographers, economists and evolutionary biologists alike recognize the importance of understanding this virtuous cycle.

Kids are getting dearer

Until the late eighteenth century, everybody either subsisted by hunting and gathering, herding, gardening or subsistence farming, or they made a living in an economy powered by manual labor and draft animals. In the last 250 years or so the industrialization of agriculture, manufacture and transport has wholly transformed the lives of people, first in the United Kingdom, then in Europe and North America and after that in pockets around the world. Few changes in human history were as rapid, as sudden, or as far-reaching as the Industrial Revolution. Over more than a century, economies and the incomes of people working in those economies grew faster than ever before. Yet for the vast majority of people today, the Industrial Revolution remains something that happened to somebody else.

In some places, like northern Ghana and much of the rest of Africa, people's livelihoods still depend mostly on their labor and livestock, and depend very little on industrial technologies. In China and India, the world's two most populous nations, the changes that typically accompany industrialization are not yet complete. The modernization of these two giant nations and their economies, and the transformation of the lives of their citizens is probably the most important thing happening in the world right now.

Industrialization induces many changes: children stop contributing to household income; the costs of raising and educating each child escalate; childhood mortality abates; women enjoy more education, economic opportunities and political power; and modern contraception technology and abortion become available. Every one of these developments can reduce birth rates. Every society experiences some or all of these changes as it industrializes, usually far more rapidly than the British and Europeans experienced them during the Industrial Revolution.

In pre-industrial Britain and Europe, growing populations meant an ever-smaller proportion of the population could live on the land.

When an agricultural society uses up most of the land that is suitable for growing crops and grazing livestock, the added unskilled labor that each new child brings to the family is no longer as valuable as it was, erasing the benefit of producing large numbers of offspring. The productivity of the family hits a ceiling imposed by the amount of land and the technology available with which to farm it. When children grow up, they must share the same area of land with their siblings, or some of them must find alternative ways of making a living such as joining the military, taking holy orders, or learning a specialist trade. Dividing up the land among the offspring leads quickly to poverty and, often, to evolutionary oblivion. That is why, in many societies, families that are hemmed in and unable to expand into new farming areas have invented protocols to pass land only to one of the offspring.

The saturation of farming land compelled parents to have fewer children. But not too few; as the Industrial Revolution began, child labor remained important. Where once children helped out on the farm, they became an important feature of factory life early in the Industrial Revolution, as they are in some developing economies today. Child labor can bring the family much-needed income, giving parents some remaining incentive for having large families. Regulation of child labor such as the English Factories Act of 1833, which restricted children under nine from working, limited the hours that children could work and imposed mandatory schooling for two hours a day, effectively raised the cost of child labor, thus reducing demand. It also dampened the incentives for families to send children to work in the factories.

As the Industrial Revolution took hold, the world of knowledge and the value of education were being transformed. Families that could afford to educate their children reaped tidy benefits since it was educated people who drove the industrialization of agriculture and industrial production. Education became the surest way to ensure their social mobility. Lavishing plenty of time and effort on a child, and giving him or her the best possible education, gives that child a head start in the competition among peers that will decide who gets the best jobs, makes the most money and finds the best mates.

Similarly, in countries that are currently industrializing, the benefit to a family of having lots of cheap unskilled labor plummets when child labor is outlawed. Instead, families that invest in the formal education of their children see dramatic returns on that investment when those children grow up, earn high wages in the changing economy and marry into other upwardly mobile families. In order to afford these investments, most families need to have fewer children. In short, the best strategy of reproductive investment changes from producing many to producing fewer children and investing more heavily in each of them.

During the Industrial Revolution infants and children also became far less likely to die than ever before. Natural selection has long favored parents who produced slightly more children than they could comfortably raise because some of those children were likely to die young. A family that produces few children who all die in infancy is an evolutionary dead end. For parents it is costly to produce more children than they can afford to raise, but it is nowhere near as costly as leaving no descendants at all. As a result, when there is a high risk of children dying before they reach reproductive age, the best evolutionary strategy is to produce slightly more babies than one could hope to comfortably feed, clothe and teach, knowing that they are unlikely to all survive.

Over millions of years, natural selection has fashioned parents to hedge their reproductive bets more heavily when child mortality is high than when it is low. Since the Industrial Revolution began, families have escaped from starvation, hygiene and sanitation have improved, and medical advances like immunization and antibiotics have made decisive inroads on infectious disease. Each of these changes improved childhood health and survival, and the balance between the cost of producing too many babies and the chance of losing children shifted decisively. The optimum family size rapidly shrank as insurance against childhood mortality grew less important.

If mom had her way

Everything I have said so far about reproductive decisions assumes that families are happy units, and that what is good for mom is also good for dad. But families are riven by conflicting evolutionary interests. The conflict that starts at the top, with mom and dad, is one of the main concepts in chapters 5 to 7, so I will keep my description very brief here.

Within a family, men and women share largely the same evolutionary and economic interests. For the most part parents work together for the benefit of one another and, especially, their children. But men and women have one night stands and affairs, get divorced and sometimes one partner dies. So although a couple share a strong evolutionary interest through their children, their interests are not identical because each partner could, in the future, have children with other mates. When a woman dies in childbirth she loses the chance to have any more children or to help her children and grandchildren grow up. Her husband also loses the contribution she would have made as a mother and grandmother, but devastating though losing her might be, he could always take a new wife and start again.

Every baby is much more costly for a mom than it is for a dad. Not only have women always run a very real risk of dying in childbirth, but their bodies carry the heavy burden of pregnancy and breastfeeding. In evolutionary terms, women have more to gain than men do by having fewer children, and investing more in the upbringing of each child and grandchild. These differences in men's and women's evolutionary interests set up an evolutionary power struggle.

The effects of sexual conflict are often subtle and are anything but deterministic; we all know couples in which the woman wants more children than the man. But in the most general terms, men have long been selected to favor bigger families than women. Men can be indifferent or less averse than their partners to the next pregnancy. Charles Darwin's beloved wife, Emma, recorded her menstrual periods in her diary, and was clearly devastated at the news of some of her later pregnancies. Charles was not as keenly devastated. If Emma had had access to twenty-first century contraception, we can be certain she would have borne fewer than ten children.

Sexual conflict is a basic and powerful evolutionary force, yet it hides in such plain view that it often cannot be seen—especially by men, in whose favor most societies have tipped the balance most of the time. But sexual conflict is also so pervasive and so powerful that a shift in the balance of power can have sudden and profound consequences. By far the most important determinant of birth rate in a modern society is the status of its women. Fertility plummets when societies educate girls, grant women the vote, remove restrictions on the kinds of work women can do and allow women to own and inherit property. Supporting women's participation in the workplace with maternity leave and accessible childcare, and ensuring women have access to cheap and safe contraception and abortion, can restore to women the capacity to manage their own reproductive futures. Simply by doing this, women will have far fewer children than they do when men hold all the cards. Here there is a remarkable convergence between evolutionary biology and feminism. As Germaine Greer observed 40 years ago, "The management of fertility is one of the most important functions of adulthood."

A feminist revolution always helps, but societies don't strictly need one in order to improve the status of women. They can start by improving girls' education. Every year of education that a girl receives reduces the number of babies she will have in her lifetime. Today, in fast-growing countries like Pakistan, Nigeria and Yemen, a woman with no education produces about twice as many babies in her lifetime as a woman who has completed secondary school. Women who have completed only primary school fall somewhere in between. Recent events in Iran illustrate this dose-dependent effect of education on fertility.

Fundamentalist religion is often opposed to women's education, political empowerment and economic opportunities. Fundamentalist

Islam certainly has that reputation. One need only think of the psychopathically sexist Taliban in Afghanistan, who banned girls from attending school. As a result only 32 percent of young Afghan women can read, compared with 83 percent of young men. Women in neighboring Iran endure some of the most stultifying strictures in the Muslim world. As I write this, Sakineh Mohammadi Ashtiani, a woman from the north Iranian city of Tabriz, had received 99 lashes and was sentenced to be stoned to death for adultery. This kind of savage misogyny permeates the fabric of contemporary Iran, yet there is also a limited success story to be told. Around the time of the 1979 revolution, Iranian women had an average of seven children in their lifetime. Fertility rose throughout the 1980s owing to government policy that encouraged reproduction and did not support family planning. In 1988, after the war with Iraq had ended—a war that killed or wounded over a million Iranians—population experts convinced the ayatollahs that rapid population growth was not in the country's interests. The leadership implemented family planning policies and emphasized the education of girls. Just before the revolution, the 1976 census shows that only 10 percent of rural women aged 20–24 were literate. By 2006 this number stood at 91 percent. Iran now has the best-educated population and the greatest gender equity in education in the Middle East. It also boasts the most dramatic recorded fertility decline in any country: in 2006 the average lifetime number of children had plummeted to 1.9 per woman.

Education delivers women greater power within society and within the family for several reasons. Educated women can better avoid being coerced into having more babies than they would like. Educated women are also more likely to enter the workforce, and once there they earn better wages than less-educated working women and, sometimes, men. Those wages become an important part of family income, and the time mothers spend out of the workforce hits the family budget harder. In the language of economics, it presents an opportunity cost. As anybody trying to raise a family in the twenty-first century knows, this cost is not an arcane theoretic quantity: the time spent without the mother's wages (or the father's wages in those families where dad stays home and mom goes back to work) is a very real cost. And, for most of us, erasing this opportunity cost by having both parents go back to work creates an enormous extra cost: daycare.

It may have long been in men's evolutionary interests for women to stay home, raise lots of kids and perform the unpaid domestic work we euphemize as "home making." But modern men are finding that pairing up and raising a family with an educated woman who can earn a good wage as a more equal collaborator can be in those men's economic interests. The fertility drop in the demographic transition is driven by a virtuous cycle; the status of women improves, leading to greater participation of women in the workforce; the greater supply of labor stimulates economic growth, which in turn reinforces the need for education; this elevates the parental investment needed to raise a child; and educated mothers are more likely to ensure that their sons and especially their daughters get educated.

From an evolutionary point of view, shifting the balance of power within families and societies toward women leads to a reproductive strategy much closer to the evolved female optimum than was typical in our subsistence-farming past. Women, given control, prefer an intermediate number of children, lavishing more care on each of them. This shift can start purely from improvements in girls' education, as it has in Iran, but it is helped along by access to family planning services, contraception and abortion. Most babies today are born in the poorest nations. A great many of these babies, particularly in sub-Saharan Africa, are unplanned or unwanted by the mother. Many mothers would like to delay having their next child until they can withstand the physical rigors of pregnancy and motherhood and when they have the means to raise a child. Not only would planned parenthood suit these individual mothers, but in so doing it would dramatically curb population growth in the fastest-growing countries in the world.

About a quarter of the 120 million or so pregnancies that occur each year in the developing world are unwanted, and 4 million or so of these end in abortions. For many women, especially poor women, in these developing countries, illegal abortion is the only way to avoid complete physical, economic or social ruin. These abortions are unsafe and too often deadly or debilitating. The policy of several governments and agencies, including the United States under Reagan and the Bushes, and the Catholic Church, not to provide family planning aid to groups that perform abortions or counsel women on how to obtain an abortion is doubly destructive. Not only does it put safe abortion out of reach of the desperate women who most need it, pushing them toward unsafe illegal operators, but it also destroys an opportunity to reach the women who need other family planning services most. All in the name of religion.

Combining evolutionary life history theory with some basic home economics shows that Hardin's tragedy of the commons is not inevitable, at least when it comes to population growth. People can act in their own economic and evolutionary best interests by regulating the number of children they have, especially as education becomes the pathway to success and social mobility, and as empowerment allows women to control their reproductive destiny. Concluding *The Means of Reproduction: Sex, Power and the Future of the World*, Michelle Goldberg points out that we should not need the prospect of Malthusian misery, in all its iffyness, to make governments and organizations take women's needs seriously. Those needs are important purely because individual women are important. I hope that this chapter has eked out some small additional insights from evolutionary biology to help to explain why, in Goldberg's words, "there is no force for good as powerful as the liberation of women."

5

Shakespearean love

There are more important things than love. Like friendship. And irony. Irony is the most important.

—Christopher Hitchens, 2010

Natural selection is all about reproduction. For humans, as for other organisms that reproduce sexually, it takes two to tango. As a background to the next five chapters, each of which is soaked in sex, I provide a primer on sex, love and the issues that concern those of us who think about sex for a living. I place human sex and reproduction in the context of the remarkable carnal variety of the animal world, and the tug-of-war between cooperation and conflict at the heart of sex. I also discuss how our emotions, especially love, are part of the mechanism by which our sexual behavior evolves.

Let me not to the marriage of true minds
  Admit impediments. Love is not love
Which alters when it alteration finds,
  Or bends with the remover to remove:
O no! It is an ever-fixed mark
  That looks on tempests and is never shaken;

It is the star to every wandering bark,
Whose worth's unknown, although his height be taken.
Love's not Time's fool, though rosy lips and cheeks
Within his bending sickle's compass come:
Love alters not with his brief hours and weeks,
But bears it out even to the edge of doom.
If this be error and upon me proved,
I never writ, nor no man ever loved.

SHAKESPEARE'S SONNET 116 IS AS profound an expression of eternal and unflinching love as ever was writ, exquisitely capturing the feeling that the love between two people is absolute and permanent. Romantic love is one of the most profound emotions in our human repertoire. Its mysterious power routinely transforms lives, motivates great acts and inspires sublime creativity. Yet the ultimate evolutionary reason humans have the capacity for such potent love is something so common that we take it for granted: baby making. It takes a very special form of persuasion for two independent and unrelated individuals to come together and stay together despite all their many differences for long enough to breed. Love enables this to happen, tricking couples—sometimes only fleetingly—into feeling as though they are "no more twain, but one flesh."

But when we fall in love, will it last forever? Dr. Samuel Johnson famously described second marriages as "the triumph of hope over experience," yet even at a couple's *first* wedding we often have to suspend in hope our disbelief that anybody can love as fully or as profoundly as the wedding rites demand. Experience and ample evidence suggest that far too often love *is* time's fool, and in the matter of unflinching love Shakespeare was more wrong than right. Love is many things, most of them deeply wonderful, yet it is a rare couple for whom it is an ever-fixed mark. Many couples who are truly, madly, deeply, irritatingly in love find themselves in time all out of love.

Love inspires so much great literature, music and art, yet if affairs between men and women were truly governed by Sonnet 116, both art and life would be a whole lot less interesting. We would not have Othello's jealous wrath, Hamlet's confused paralysis, the witty bickering in *The Taming of the Shrew*, or the tragic star-crossed lovers Romeo and Juliet, whose kinship ties to warring clans seal their fate. To put it another way, if music be the food of love then we are only partly nourished by torch songs, tender ballads and love songs. Nothing can touch —let alone wash away—the blues.

In the coming chapters I explore some of the deep-seated evolutionary reasons that affairs of the heart fascinate us so wickedly. At first glance, it is staggering just how much courtship practices, marriage customs and family arrangements vary among cultures and among times. Yet this variation is miniscule when viewed from the perspective of a biologist used to considering the variety of animal mating systems. Importantly, the main features of human mating systems and the ways in which economic and cultural factors modify those mating systems are predictable and understandable when viewed in the context of the evolutionary forces that shape animal mating systems. In order for you to understand what I mean, I should explain a few things about how typical mammals reproduce, and how in some ways humans are more like typical birds than mammals.

The importance of being a mammal

Every person alive today is the descendant of a long line of successful ancestors. In the 5 million years since we last shared an ancestor with chimpanzees, we have each had roughly 200,000 pairs of ancestors who successfully found one another, mated, negotiated pregnancy and breastfeeding and raised one or more children. This history of successful ancestors goes back well beyond 5 million years, at least as far as the very first single-celled sexual organisms well over a billion years ago.

Remember that at the same time as our ancestors were mating, giving birth and raising families, many of their contemporaries died too young to mate, never found a mate or left no descendants who grew up to reproduce. The one thing we know about our ancestors is that, reproductively at least, they were each among the most successful individuals of their generation. The cumulative power of all this sex and reproduction is that we are fine-tuned to turn just about any circumstance to our reproductive advantage.

In evolution, living to a ripe old age is only worthwhile if it means you get to have more babies and raise them to have babies of their own. Some of the strongest selection known to science occurs because the individuals who most effectively find, compete for, seduce or even coerce suitable mates pass on to their offspring the genes that will enable those offspring to do the same. Success in finding and winning mates leads to a special form of natural selection that Charles Darwin called sexual selection.

Although each of us and each of our ancestors had a mother and a father, the most evolutionarily successful women and men in history achieved their success in very different ways. Sexual selection acts differently and favors somewhat different strategies in males and females of most species, including men and women. Sexual selection has shaped the bodies, brains and behaviors of men and women in ways that are often quite different, and some of these differences come about in typically mammalian fashion.

For a female mammal, motherhood begins with the massive investment of carrying the developing young in her uterus and nourishing them through a placenta until they are born. Her body then produces milk that contains all the energy and nutrients the growing young need until they can be weaned onto solid food. She has to travel with the young at a slow enough speed for them to keep up with her, or return to the young several times a day to feed them. As a result she is heavily constrained in how much food she can obtain for herself and her young, and both she and the young are at serious risk of being killed by predators. For some mammalian mothers, like mice, things return to normal once they wean their pups, who then move away forever.

In other species, mothers spend years teaching the offspring everything they can, and they often have offspring of several ages together with them in the social group. Elephant cows live in extended herds often comprising their mother, sisters, adult daughters and nieces as well as their immature sons and nephews. It takes more than a decade to raise an elephant to the point at which she knows enough to have her own calves. The most successful mammalian females tend to be those that can gain the resources they need in order to bear the enormous burden of pregnancy and milk production. The huge investment that mothers make in each offspring makes every female incredibly valuable to males as a mate.

Mammalian fathers, on the other hand, typically get low marks for effort, tending not to have much direct involvement in the raising of offspring. Mothers mate with the male who holds ownership of the place or the social group in which she lives, and the key to male success is winning dominion over such a place or group and defending it from would-be usurpers. Elephant seal males of the Southern Ocean come ashore at a few particular beaches in September and fight viciously to achieve dominance over one another, knowing that females will shortly come ashore to pup. After each female gives birth, a dominant male then mates with her. But his evolutionary success comes at an enormous cost because of the injuries he sustains and the energy he loses in a never-ending battle to maintain his dominance over other males.

Evolutionary fitness for males across the entire animal kingdom, and especially for mammals, tends to be a boom or bust affair. Very few male elephant seals grow big enough, strong enough, and have sufficiently lucky timing to ever become a beachmaster. Most are killed or maimed in fighting along the way and die before they ever have a chance to mate, leaving no offspring. These males are evolutionary dead-ends and their particular combination of genes dies with them. The few that do become beachmasters—if only for a few weeks—tend to hit the evolutionary jackpot. These are the males who sire all of the sons and daughters of the next generation, and every elephant seal that is alive today can trace its ancestry back through a lineage of successful males like this.

Because only the biggest males can dominate a beach, any genes that make males bigger tend to be inherited by his many offspring. As a result of this sexual selection, male southern elephant seals have evolved to be up to eight times heavier than a mature female. In most mammals, body size can be the best weapon, and mammalian males are often much larger than females of their own species. Other weapons in the fight for dominance over other males—like the horns of the impala and the teeth of the lion—have evolved to their current exaggerated size via thousands of generations of sexual selection in which longer-horned or sharper-toothed males have vanquished their less well-endowed rivals in bloody combat.

Even though impala, lions and elephant seals are in many ways typical mammals, every one of the 4500 species of mammal alive today differs in mating behavior from every other. There is even a lot of variation among our closest living relatives. Gorillas are the most typically mammalian, with a silverback male defending a harem of around five females and mating exclusively with them all for the four to five years before he is rolled by some young upstart.

Chimps and bonobos, our closest living relatives, are different. Troops comprise several mature males, mature females and their offspring. Fertile females mate with several males in a cycle and there is not very much male–male fighting over access to females. As a result, chimp males and females are roughly the same size. Humans are somewhere in between chimps and gorillas. Men are usually about 10 percent taller than women and considerably more muscular, suggesting that physical fighting and territoriality among men have been an important part of our mating system for at least a few million years.

A lot like birds

If our violent mammalian heritage paints a bleak picture of human reproduction, it may be comforting to know that we are rather unusual mammals. In a lot of ways we are more like birds. Chicks of most bird species are difficult to raise. There is no warm cozy uterus in which to develop so the eggs must be kept warm until the chicks hatch, and this usually means one parent sitting on the nest at any time in order to incubate the eggs. Chicks have to grow as quickly as possible from helpless hatchling to competent pilot and then to total independence. If you have ever had the pleasure of watching a pair of birds nesting in your garden, you might have noticed just how busy both parents are. It can take the frantic full-time effort of two parents, sometimes even with helpers, to feed a nestful of helpless chicks.

Female birds won't mate with any old cock. They tend to prefer a mate who can show that he is healthy and committed to helping her care for the chicks. In many species, females choose to mate with a male who has built a nest that is safe from predators or close to abundant food sources. In species like albatrosses and eagles, in which it takes everything that both parents can muster in order to rear just one chick through to fledging, once a pair have chosen one another they tend to stick together, mating with one another each year for decades.

Although birds were long thought to be paragons of fidelity, advances in molecular genetics tell a story of hidden betrayal. In many bird species, once the female has found, shacked up and mated with the male who owns the best territory or nest on offer and who shows all the signs of being an attentive father and mate, her eye starts to wander. DNA analyses show that in 90 percent of bird species some of the chicks that a female rears with her social mate are not fertilized by him. In those species, 11 percent of chicks are sired by another male, on average. But the Australian superb fairy wren holds the world record with 95 percent of females fertilizing some eggs with the sperm obtained from another, sexier, male in a premeditated but very sneaky predawn visit to his nest.

Female birds can make two distinct types of mate choice: choosing who to nest and co-parent with, and choosing whether to mate with one or more males outside of that pair bond (biologists call this an extra-pair copulation). The first type of choice secures the resources and help that she needs to successfully fledge her chicks. Most males who secure a good nest site get to mate and rear a family with a female in this way. When a female mates extra-pair, she has sex with a particularly attractive male who can show he has a fine combination of genes. She is also much more likely to stray like this if the male she has nested with shows signs of being genetically below par. The female seldom gets extra offspring from her infidelity: instead her offspring benefit because the genes of the extra-pair male make his offspring with her better adapted to survive and reproduce in turn. Like a good fund manager, she hedges the most important investment she will ever make: her chicks.

Males are seldom as stupid as they look. If a male bird detects any hint of infidelity in his partner—even a suspicious absence during the critical egg-laying period—he puts less effort into rearing the brood. Sometimes a suspicious dad will even abandon the female to raise the chicks on her own. After all, it makes little evolutionary sense to spend time and effort rearing some other male's chicks. However, it takes two to tango, and males also do their bit to undermine monogamy. A male who can score extra-pair copulations with females wins a massive fitness payoff: more offspring without the hard work of incubating the eggs and rearing the chicks.

Both forms of female mate choice (choosing a social mate to settle down with and deciding whether to get a bit on the side and with whom) create very strong sexual selection on male birds. Instead of competing physically with one another for dominance, most male birds compete by showing off to females in the hopes of persuading the females of the quality of paternal care they can offer, the quality of their genes, or both. This explains why male birds are the biggest show-offs in the animal kingdom whereas female birds are so often dull. The spectacular tail feathers of male birds of paradise, the calls of male lyrebirds and the flashy iridescent blue of superb fairy wrens are all the evolved result of strong sexual selection in which females judge the showiest males to be the most worthy of mating.

Human children, like chicks, are also very difficult to raise, requiring a decade or two of hard work from one or both parents. The effort that a mother and father spend raising the children constitutes a valuable commitment, not only to the children but also to one another. As a result, both men and women tend to be quite fussy about whom they pair up with to mate and raise children. Mate choice is important in all human cultures. The choice of marriage partner is an important form of mutual mate choice between men and women in most societies, and it is quite common for men and women to marry several times in a lifetime, exercising mate choice each time. Even in those societies where families arrange marriages to suit economic interests and forge political alliances, sexual selection drives the elaboration of traits that make somebody a valuable marriage partner, including good looks, intelligence and the desire to get ahead.

Just like in birds, men and women also exercise a clandestine yet powerful form of mate choice when they have sex with somebody other than their spouse or long-term partner. One or both partners may be surreptitiously—even consciously—looking for a new long-term partner. In most cases, however, the evolutionary rewards that men and women gain from extramarital sex differ somewhat. For a man, any mating is a chance to add to the number of children he sires. Within a marriage his reproductive success is limited by the number of babies his wife can bear him, so in evolutionary terms a short-term mating is like winning the lottery. Every one-night stand can potentially beget a child that he doesn't have to raise.

A woman who mates outside her marriage might have more babies only if her husband cannot fertilize her eggs. Women can, however, improve their children's prospects by mating with a man who has an excellent combination of genes because the children will inherit (half of) those genes rather than the more mediocre genes of her husband. Women—both married and single—can also often secure resources for themselves or their children by mating discreetly with men who have access to those resources.

What I think about when I think about sex

In England, the Victorian era (1837–1901) delivered explosive progress in technology and agriculture, transcendent changes in art and literature and profound growth in rational and progressive thought. The foundations of modern utilitarianism, feminism, socialism, and modern parliamentary democracy were laid in Victorian England. Darwin's great works on evolution were an intellectual highlight of the era, but by no means an isolated one. Nonetheless it is not a time that is known for an equally freethinking and liberated attitude to sex.

Today, Queen Victoria and the era her reign defined are synonymous with prudishness and sexual repression. Although the working class had a far more liberated attitude to sex, with one-third of brides estimated to be pregnant on their wedding day, the upper classes were particularly stiff. Courtship was rigidly orchestrated and chaperoned, involving little more contact than hand-holding. While the resulting sexual tension created a niche for great literature, it also fostered a repugnantly prudish ignorance.

In 1857, Dr. William Acton published an immensely popular book on reproductive anatomy and sex, which fueled the hysterical beliefs of the time that masturbation led to the degeneration of both mind and body, including blindness. He also almost completely failed to mention women other than to observe that "the majority of women (happily for society) are not very much troubled with sexual feelings of any kind."*

In many ways biology has been carrying Victorian baggage for the last 150 years, and we have only recently entered our own sexual revolution. Although Darwin's books transformed scientific thinking about reproduction, the delicate and sanitized ways in which his fellow gentlemen naturalists thought about sex are still with us today, shaping and distorting the ways in which many people understand reproduction. Even today, the euphemistic view that sex is a necessary act that happens for the "perpetuation of the species" prevails in many scientific papers and nature documentaries alike. Others take the sanitized view that sex is a happy and cooperative event, best discussed discreetly, seen in soft-focus and not too close-up.

Most people either ignore or do not realize that males and females maximize their evolutionary fitness in very different ways. For most animals, mating involves one part cooperation and several parts exploitation, and this is true in humans too. Basically, when a male and a female mate, they each need to get as much out of the other as they can, for their own benefit as much as for the benefit of their mutual offspring. What is good for the goose is very literally not always good for the gander.

The vinegar fly *Drosophila melanogaster*, commonly known as the fruit fly, is a favorite study organism for evolutionary biologists. Because they take only two weeks to go from egg to adult, a scientist can introduce populations of flies to a new set of environmental conditions, and within a year or two get a very good picture of how those conditions changed the direction of evolution. In the late 1990s, Brett Holland and his PhD supervisor, Bill Rice, at the University of California at Santa Cruz audaciously demonstrated the full power of sexual conflict in an experiment on these flies. Instead of changing the temperature, or diet, or humidity to which

* William Acton's example illustrates that a male author can write about human sex with only limited insight and authority, a lesson I will try to keep foremost as I try to wring this chapter from my word processor.

flies were exposed, they changed the entire mating system. Female vinegar flies lay their eggs in rotting fruit, and males hover about pieces of rotting fruit waiting for a female to come by. Males compete vigorously for the chance to mate with any female, and a female tends to mate with several males before laying her eggs. Holland and Rice let this competition and multiple mating happen in three experimental populations; at mating time, each female was placed in her own vial with five males.

The simple stroke of genius is that Holland and Rice forced another three populations of flies to be monogamous; at mating time each female was placed in a vial with only one male and left to mate and lay eggs. If a male or female died, they were never replaced. The scientists recognized that by enforcing strict monogamy for the admittedly brief term of the flies' reproductive lives, they artificially made the interests of the male and the female identical. Anything the male did that harmed his mate also harmed his own reproductive success, and the same was true of the female. As a result, after more than two years, and 50 fly generations of evolution, the monogamous fly lines had lost much of the tendency for males and females to harm one another. Mating with a male from a monogamous line did not result in the same reduction in female life span as mating with a male from a polygamous line. Males from monogamous lines also did not try to copulate as often as flies from the multiple mated lines.

Flies from the multiple mated lines were so nasty because the reproductive interests of males and females were very different. For a male, the main game was to mate more often than his rivals, and ensure that females lay as many eggs as possible using his sperm. Male flies have evolved, over millions of years, any number of dirty tricks to outcompete rival males. They ejaculate, combined with their sperm, a cocktail of chemicals that cause females to lay more eggs than they would have liked, and that give the male's sperm an edge over others' sperm in the race to fertilize those eggs. These chemicals often directly harm the female, shortening her life and reducing the total number of eggs she lays in her lifetime and thus her evolutionary fitness. Male flies with gentler semen fail to fertilize as many eggs, and so any genes that cause gentle semen reach the end of the evolutionary road.

This is more than a tale of nasty males. Females from multiple mated lines have stronger defenses, both physical and chemical, to prevent males from mating and to minimize the harm caused by the nasty chemicals in males' semen. In the lines where monogamy was imposed, not only did males evolve to be less arduous and less toxic, but females also lost most of their tendency to resist mating attempts from males, and their resistance to the chemicals of nasty males also started to slide.

Interestingly, the population does better under monogamy. Without the need for males to compete and to manipulate female reproduction, and without the need for females to resist the incessant advances and toxic semen of males, females lay more eggs in total and so the population grows faster. As soon as there is the chance of the female mating with two males, it no longer pays the male to be nice. The males who court most persistently, cajoling or coercing the female into mating most often, and whose sperm are most effective, end up fertilizing the most eggs. This ensures the success of nasty genes. Male and female flies are engaged in a vicious evolutionary arms race that ends up, like all arms races evolutionary and political, dragging both down. Once again, the idea of reproduction maximizing the benefit to the species is shown to be fly-blown garbage.

It might be tempting to moralize, directly from the experimental evidence in vinegar flies, that imposing lifelong monogamy on humans would solve many of our greatest social problems by eliminating the conflict of interest between mates. Unfortunately human affairs are always much more complex than the goings on in a vial of flies, and experimental social engineering is awfully unpopular on ethical grounds. Only the fictional couple marooned as virgins on an idyllic yet otherwise uninhabited island with no prospect of rescue would have identical evolutionary interests.* The rest of us have to navigate a murky world where conflicts can range from the negligible to the profound.

No sexual species of animal remains strictly and exclusively monogamous for a lifetime, and differences in evolutionary interests between males and females are the inevitable result. In every species, the interests of males and females differ, and the further the mating system strays from lifelong monogamy, this difference between what is good for females and what is good for males grows bigger. Science is only just gaining a toehold on the evolutionary conflicts between men and women, but there is ample material to study because of the huge variety in human mating systems. Societies differ in their marriage customs, in the extent to which marriages can dissolve and remarriage is allowed, and in the extent to which one or both partners can and do have sex with people outside of the union. As we will see in chapters 6 and 7, every one of these differences between groups changes the level of conflict between the interests of men and women. As a result human mating is as infested with conflict as the sordid world of vinegar flies, and the degree of sexual conflict also varies among places, times and environmental circumstances.

In heat

Monty Python famously parodied the Catholic Church's antediluvian position on contraception with the musical sketch "Every Sperm Is Sacred." Far from being sacred, to a biologist each sperm cell is almost entirely insignificant. By that I mean sperm are made small and fast because the second each sperm leaves the male body it is thrust into a frantic race to find and fertilize an egg. A male invests only tiny amounts of energy and resources into making each single sperm. He makes billions of them, yet it is a single sperm that eventually performs the significant—

* I am happy to act as a consultant for any ensuing reality television show.

to some, sacred—role of fertilizing the much rarer and more valuable egg. At this point both the egg and the sperm contribute the same amount of genetic information to the new individual that is conceived.

The contrasts between sperm-producing men and egg-producing women are amplified by the fact that a mother carries the baby to term through a physically and energetically demanding pregnancy and then feeds the baby on milk made by drawing on her own fat reserves. During pregnancy women cannot conceive, and this remains mostly true throughout breastfeeding too—at least in hunter-gatherers who don't have much surplus food. Men can, in theory at least, fertilize the eggs of several women in a day, and several hundred women in the time it takes one woman to carry a baby to term and wean it off breast milk.

The fact that it is possible for a man to increase his evolutionary fitness dramatically with a single anonymous quickie is a biological fact that seems to give men an unfair advantage. This is offset, however, by the fact that women have anonymous quickies too. Because eggs are fertilized inside the woman's body, a woman can be certain that she is the mother of the child no matter how quick and unmemorable the sex might have been. After all, delivering a baby, either via the traditional vaginal route or the ever more common Caesarean pathway tends to leave a bit of an impression on a mom. A dad, on the other hand, never knows with complete certainty that he really is the father. As the proverb goes: "Maternity is a matter of fact, but paternity is a matter of opinion."

Women differ from most other female mammals in one particularly interesting way: they lack a distinct period of estrus in which they advertise their fertility and to which they confine all their matings. If you have ever owned dogs you will know that bitches tend only to be sexually receptive, or in heat, for a few days at a time. They advertise this fact with pheromones—chemical signals—that permeate the neighborhood, whipping every intact dog nearby into a frenzy of carnal excitement. Most wild and domestic mammals have an estrus, during which ovulation occurs, females are receptive to mating attempts, and matings are likely to result in conception. The existence of a period of estrus solves one problem that has its roots in sexual conflict: incessant courtship and mating attempts from males.

In guppies, one of the few fishes with internal fertilization, males spend most of their day swimming about after females, showing off their gaudy color patterns in a most persistent and usually vain attempt to mate. Males also sneak up behind females and attempt to insert some sperm into the female's genital tract. It is only the males that are attractive, persistent or good at sneaking up on females that get to mate, and this means that the genes that make the most irritatingly arduous male guppies become widespread through sexual selection.

Female mammals evolved to signal estrus in order to avoid never-ending attention from males. Males have evolved the ability to detect when females are in estrus, reducing the need to fight off rivals and defend or court females year-round, and restricting these costly activities to one or a few times a year. As a result, in most mammals, estrus is mercifully brief—just long enough to ensure the female has a good chance of getting her eggs fertilized. Females don't have to weather an extended period of sexual attention and harassment and males don't have to be perennially at war with one another.

The majority of monkeys and apes do not have a strict estrus, and males and females copulate across more of the ovulatory cycle than is typical for other mammals, although sexual activity peaks at the times when females are most likely to conceive. In our surviving relatives, the common chimpanzee and especially the bonobo, females are interested in sex and seek out matings across most of the ovulatory cycle. Therefore many matings happen when there is negligible chance of conception. Females mate promiscuously, possibly to ensure that each male in a female's troop has reason to believe that her next baby could be his. Males are jealous and dangerous things, and thousands of generations of selection have tuned males to detect babies that could not possibly be their offspring. Such babies can be abused or even killed.

Chimp promiscuity may be a female strategy to muddy the waters of paternity in the interests of her offspring, but it only works if there is no way for males to know when the female is at her most fertile. Chimp females advertise when they could be fertile; the skin around their genitals becomes bright pink or red and often swells into protuberances that advertise her sexual readiness and possible fertility. But this advertisement starts well before and lasts until well beyond the period during which she can conceive, and that might be enough to fool the males.

Humans do not grow sexual swellings, pump out pheromones, or in any other obvious way advertise when they are near the most fertile phase in their cycles. In the time since we last shared a common ancestor with the chimps, humans have taken the chimp-like trait of copulating outside of the most fertile period to new extremes, in which we can and do mate at just about any time, and men are usually none the wiser about whether a given copulation could be the lucky one.

But why have women evolved to be so very cryptic about when they are fertile? Both chimps and humans inherited the tendency to have sex when females are not fertile from our common ancestor, an ape that was probably promiscuous much like chimps are today. Concealed ovulation, as we call it, might be a female device to confuse paternity, ensuring that several men have an interest in protecting and contributing materially to raising a woman's children. Concealed ovulation almost certainly first evolved in our more promiscuous ancestors to keep males guessing about their paternity, but that might not be the whole story.

Concealed ovulation could, instead, be a kind of modesty designed to minimize the jealousy of the woman's long-term partner and harassment or coercion from other men. Imagine the mayhem in your neighborhood or workplace if women advertised their most fertile and sexually receptive periods and if men could detect when women were ovulating. Women would receive incessant attention from their partners, male acquaintances and strangers for a few days a month. Husbands would probably not move out of sight of their wives near to ovulation for fear that their wives would have sex with the first moderately shaggable stranger that came along. Women would become an even bigger source of conflict among men than they already are, potentially resulting in injury or death to the woman, the children or her husband.

But this dangerous period would only happen at peak fertility. Men —be they partners or possible lovers—would pay far less attention to a woman's material and emotional needs at other times of the month. One can easily see how women who keep men guessing about ovulation would benefit, both by keeping them interested throughout the month and by toning down male jealousy and harassment when they are most fertile. Concealed ovulation appears to be a triumph of women's interests over those of men, and it might be one of the reasons human societies don't degenerate into sexually frenzied chaos.

But concealed ovulation is unlikely to be stable because any man with the capacity to detect when a woman is fertile would enjoy an enormous evolutionary advantage over other men. Husbands would know when to keep a close watch on their wives, and men looking for one night of passion would know exactly whom to pay the closest attention. It seems an essential failing of the human male that very few of us can recognize a woman who is not only fertile, but sexually interested in us, despite the obvious advantages we would enjoy if we were better at it. I optimistically put this down to the sophistication with which women conceal their interest and their fertility from us. Yet there is a growing body of evidence that men do possess some capacity to detect when a woman is at her most fertile, and act accordingly.

My favorite example, for its sheer audacity and originality, is an experiment by Geoffrey Miller and his students Joshua Tybur and Brent Jordan, who enlisted 18 women working in strip clubs in Albuquerque, New Mexico. Strippers in these clubs make most of their money as tips for performing lap dances; the more lap dances they can persuade men to "purchase," the better their earnings for the night. The women who agreed to participate in the study submitted data on their earnings per shift as well as their menstrual cycles. Strippers at the most fertile part of their cycle made approximately $335 per five-hour shift, whereas they only made $185 per shift when they were menstruating and $260 per shift in between their periods and their fertile peak. Interestingly, the seven women who were on the contraceptive pill, which fakes the physiological conditions of early pregnancy, made about $50 to $100 less per shift than naturally cycling women made, and their earnings were much more consistent across the month.

The precise mechanisms behind this result remain unclear—a subject ripe for further careful yet sensitive research. Were the women more attractive at their fertile peak because they exuded some verbal or visual sign that men picked up on? Or were they just better at exploiting the men's vulnerabilities because the job was slightly less repellent to them when they were most fertile? It would certainly be interesting, and probably lucrative, to know more. But as is so often the case in evolutionary biology, the result remains important irrespective of the underlying mechanisms.

Miller, Tybur and Jordan won an IgNobel prize (the coveted annual awards "for research achievements that first make people laugh, and then make them think") and worldwide publicity for their paper, which, titillating as it is, reveals a serious message about human sexuality. Women are not as cryptic and men not as dumb about fertility as we once thought, and the dynamics that operate so acutely in strip clubs probably permeate our workplaces and social milieus in more subtle ways. A growing body of work shows that women, without conscious forethought, change their behaviour during the most fertile part of their cycle, and that the contraceptive pill tends to blunt these changes. The ability of women to conceal ovulation and fertility, and the capacity of men to detect very subtle changes in women's behaviour, are traits fashioned by sexual conflict. There can be no winner in sexual conflict, merely ongoing cycles in which the evolution of an advantage to one sex selects ever more strongly on the other sex to thwart that advantage.

Love is the drug

How do we square this conflict-riddled view of sex and mating with the rose-tinted Shakespearean wonder of falling in love? Most of us remain convinced that feelings like love, lust and jealousy govern our actions, and the idea that we might fall in love with a view to increasing our evolutionary fitness can seem like the kind of emotionally inarticulate prattling that only a stereotypic geek-scientist could come up with. But love and lust, longing and loathing, jealousy and sexual boredom are all general tools that evolution has built over time. Just as the balance between the five primary flavors and a few textural cues can tell us which of several thousand possible foods we are eating, so the interplay of emotions with our life circumstances can guide us as we navigate our complex social and sexual worlds.

Any explanation of human behavior that relies only on the feelings and motivations of which we are consciously aware is limited. There are some feelings and urges that we can recognize and articulate verbally, there are others that we can vaguely identify, and there are many others that we don't even realize are operating. Our awareness of how we feel probably evolved in order to bring our emotions and motivations, and our responses to them, under the control of the brain's executive functions. These functions are the cognitive abilities by which we anticipate the consequences of our actions, plan what we say or do, work toward a goal, and refrain from doing inappropriate things.

It seldom makes sense to declare our undying love as soon as we feel the first rush of infatuation, or to fly into a jealous rage the first time our lover looks at somebody else. We tread carefully because the consequences of misinterpreting the signals or getting our timing wrong can be catastrophic. In the same way aircraft designers put an airplane's most important controls and instruments within the pilot's easy reach and normal field of view, so evolution has brought the feelings and drives that most need our thoughtful oversight into the realm of our conscious awareness. Yet the feelings and urges of which we are aware only tell a part of the much bigger story of why we do what we do.

Natural selection molds our behavior by improving, generation after generation, the material basis of our nerves and hormones, by refining how our brains change in response to experience, and sharpening the way our organs and muscles respond to signals from our hormones and nerves. The marvel of human behavior is that we function adaptively in many millions of contexts, making us one of the most subtle and interesting creatures that ever evolved. Even more marvelous, we achieve all this complexity by varying only a limited number of component parts. In recent years scientists have gained as much knowledge about love and romantic attachment from studies of the brain and the chemical signals in our bodies as exists in all the works of literature combined.

Hormones are chemical signals that can send the same message to several different parts of the body. Adrenaline, for example, is released when we face immediate danger. When adrenaline spikes in the bloodstream, we get the familiar rush in which our heartbeat quickens, the airways in our lungs open up and our blood vessels dilate. All of these responses ensure that plenty of oxygen is available to the muscles so that we are ready to fight or to flee the situation. The reason all of these different body parts respond to adrenaline is that their tissues are infested with receptor molecules that bind to adrenaline, changing the properties of the cells and thus triggering a response in those cells. The more receptors on the surface of a cell, the more dramatic the response to adrenaline will be.

Chemical signals are seldom as simple as the adrenaline pathway. They usually involve several hormones and neurotransmitters (chemical signals that transmit information among nerves), and body and brain tissues that respond to each chemical in different ways. At the heart of love lie at least four such chemicals: dopamine, serotonin, oxytocin, and vasopressin. Just as the ingredients in a good meal work together to stimulate our senses, so these chemicals do not operate in isolation from one another. But to keep things simple I will pay some attention to each of these chemicals one at a time.

Dopamine is an astonishingly versatile neurotransmitter that is involved, among other things, in the brain's systems of motivation, desire and reward. Modern brain-imaging technologies show that when people look at pictures of their lovers the most active parts of the brain are areas especially sensitive to dopamine. Dopamine is probably involved in the wonderfully euphoric feeling we get when we are falling in love, rewarding us as we bond with somebody that might be a good partner. Dopamine is also involved in the euphoria of orgasm. Think of this orgasmic high as an evolved reward for having sex. The dopamine-mediated processes in the brain that regulate pair-bonding behavior and love are also crucial in the development of drug addiction. Drugs like cocaine and amphetamines, and quite possibly alcohol, nicotine and heroin appear to co-opt some of the reward pathways the brain uses to form and reinforce bonds with lovers and friends.

Falling in love also brings a drop in the neurotransmitter serotonin, an important signal controlling feeding, growth and reproduction. Normal to high levels of serotonin are essential for a feeling of well-being. Drugs like Prozac that block the uptake of serotonin into cells, leaving more serotonin available in the brain, are important tools in treating depression, obsessive-compulsive disorder and anxiety disorders. People in the early stages of romance can be quite literally love-sick, as their serotonin levels plunge to lows typically seen in people with obsessive-compulsive disorder. It may well be that low serotonin levels, together with the euphoria that comes from a spike in dopamine, contribute to the focused and mostly happy obsession of falling for somebody. The importance of a dip in serotonin during early romance is underlined by evidence that people who take serotonin-uptake inhibiting drugs find it harder to fall in love and to bond romantically, possibly because the drugs are blunting the love-sick plunge in serotonin.*

The hormones oxytocin and vasopressin also feature in the formation of bonds and attachment between individuals. When a mother's nipples are stimulated, the pituitary gland in the brain releases oxytocin, which in turn triggers the "let-down" of milk to the nipple. The increased oxytocin (and possibly vasopressin) also enables the mother and baby to bond. If you are a mom, or have had the privilege of watching a mother get to know her newborn, you will have witnessed the formation of the closest and tightest bond in human social life. Natural selection shaped those parts of the baby's and mother's brains associated with social learning, trust, generosity and empathy to respond to the exact chemicals that peak when mother and baby are closest—the hormones involved in breastfeeding.

Despite what the most militant breast-is-best zealots might say, breastfeeding is not essential to mother-baby bonding, but it certainly does help. Mothers shown a picture of their own baby experience more activity in areas of the brain associated with oxytocin and vasopressin receptors, and stimuli other than nipple stimulation can trigger oxytocin and vasopressin secretion. Bonds between mother and child (and between father and child too) can and usually do form just fine without breastfeeding, but it seems that the massive oxytocin surge during milk let-down helps those bonds form fast.

For a scientist, one of the coolest things about oxytocin is that it can be made in a laboratory and administered via a nasal spray. This means that instead of relying on correlative evidence (for example, high oxytocin levels in the blood plasma are associated with increased trust), scientists can do experiments in which the only thing they vary is whether the subject gets a shot of oxytocin or a shot of pure saline up the nose.

* I don't mean to suggest that a blunted capacity to fall in love can in any way outweigh the necessary and important job that these drugs do in helping people fight depression and other maladies.

Subjects given oxytocin before an experiment are more trusting, generous and empathic, and less fearful. All of these effects have earned oxytocin a reputation as the "cuddle chemical," and they help us to fall in love, enabling us to overcome our natural fear and wariness, and to get close enough to an unrelated human being for long enough to mate and have children. Oxytocin release is a natural part of sexual arousal and orgasm, probably reinforcing the building of trust as well as stimulating us to be generous toward our mates—with our affection as well as our time and resources.

Vasopressin is a very similar chemical to oxytocin. Its main job is in regulating the body's water balance, but it also moonlights in regulating human social affairs. Like dopamine and oxytocin, vasopressin is associated with the brain pathways that reward us as we pair up and when we have sex. There is some evidence that vasopressin is an important element of the reward that men experience as they forge a pair bond with their mate. One reasonably common genetic variant causes the vasopressin receptor not to work properly, at least in men. Men with two defective copies of this gene are less affectionate and cooperative toward their partners, and their partners report being less satisfied in the relationship. As a result of a tiny defect in the receptor for one hormone, these men find it harder to make and sustain long-term romantic attachments.

Vasopressin and oxytocin play an essential role in pair-bonding in some other mammals, including the prairie vole—an exquisitely cute rat-like inhabitant of American grasslands. Male and female prairie voles form long-lasting monogamous pair bonds, grooming one another, sleeping together and sharing the work of raising pups. Prairie voles also occasionally have sex with mates other than their long-term partner. If researchers block the release of vasopressin and oxytocin, however, prairie voles mate promiscuously and fail to establish long-term pair bonds. The montane vole, a close relative of the prairie vole, is normally promiscuous—even if injected with large doses of oxytocin and vasopressin. Close inspection shows that receptors for these chemicals abound in the reward centers of prairie vole brains but not in the corresponding parts of montane vole brains. Monogamy evolved in prairie voles by building the hardware that rewards individuals for being (mostly) faithful partners. Human pair-bonding also has almost certainly evolved with the assistance of similar changes in oxytocin and vasopressin signaling and reception in the brain.

That was a quick and easy tour of the hormonal and nervous building blocks of love. Natural selection fashions these building blocks and their many more subtle and complex interactions in order to shape our adaptive behavior. Love is its own reward.* It is a crucial mechanism by which lovers forge special one-on-one bonds. Love is the drug that brings us together. Yet love is not the final word.

As we have seen in this chapter, while mates share an evolutionary interest in their children, their interests are also often in conflict. The love of Shakespeare's Sonnet 116 is an ephemeral treasure, a convenient and very public untruth that allows a man and woman to ignore and transcend, for a time, the differences between their evolutionary interests. We use the idea of such pure love to persuade—some might say deceive—ourselves and one another that we are doing the right thing in binding a significant portion of our life and our future reproductive potential to another individual. Yet as George Bernard Shaw observed in his 1908 play *Getting Married*:

> When two people are under the influence of the most violent,
> most insane, most delusive and most transient of passions,
> they are required to swear that they will remain in that
> excited, abnormal and exhausting condition continuously
> until death do them part.

Many people who fall in love have no intention of having children; we have evolved the hardware that equips us to love, yet often our

* This provides the near corollary to my favorite bumper sticker of all time, "Chastity is its own punishment."

circumstances or the nature of our love do not lead on to reproduction. The hormonal signals and our bodily responses evolved to add up to love precisely because they make us feel like putting aside our individuality and our conflicting interests, and once those obstacles are overcome, conceiving a child can be surprisingly easy. That people get together to raise children is one of the many everyday miracles of evolution. In the next chapter I explore the tension between cooperation and conflict that exists at the heart of this miracle, and how that tension is influenced by economic circumstances.

6

Wrapped around your finger

Marriage requires a special talent, like acting. Monogamy requires genius.

—Attributed to Warren Beatty

Human mating and child rearing involve a deep cooperation and collaboration between mother and father. But the fitness costs and benefits that the two parents experience are not identical. Understanding mating as a cooperative conflict helps us to understand why humans have such a strong pair bond, why that bond is seldom permanent, how much each parent contributes to raising the family, and the balance of power between men and women within society.

AS I BEGAN WRITING THIS chapter, in the dying days of 2009, the two biggest news stories in the world were the Copenhagen climate summit and the scandal over Tiger Woods' prolific marital infidelities. In the inane 24-hour news cycle, the complex global issue with the potential to alter the world's environmental and economic future lost out decisively to a personal tragedy that was foremost the concern of one sportsman and his family. In the words of straight-talking Australian golfer Geoff Ogilvy, "The *New York Post*'s front page had Tiger more days in a row than September 11. That's a little disturbing, don't you think?" It certainly disturbs me that the domestic lives and private betrayals of sports stars and celebrities are far more interesting to so many people than issues in which we all have a genuine stake. Yet it is beyond irony that the gossip mills—seemingly starved of content—circumvent the need for any discernible talent by manufacturing trivial celebutantes and then gushing over their domestic lives.

Before people had gossip magazines, we fueled our fascination with real gossip. About real people. The sex lives of celebrities, the fatuously famous, and real people we actually know are so spellbinding because they offer a glimpse into the push and pull between cooperation and conflict at the marrow of any sexual relationship. Anybody in a relationship has to negotiate the tension between the tight cooperation needed to build a life together and possibly raise children and the conflicting interests of mates. This tension is played out between individual men and women and within families, and is shaped by the economic and social circumstances those individuals experience. But at the level of a society, all those individual interactions sum to influence the structure of human families, how much men and women contribute to the household, how long marriages tend to last, and the relative status of men and women within the family and within society.

Reproducing sexually is the ultimate cooperative act; two individuals conspire to combine their DNA, making a new life. The importance of cooperation and commitment are reflected in the ways we venerate romantic generosity, kindness, sexual fidelity, and selfless parenting—all traits that reinforce the cooperative bedrock of sex and reproduction. Yet the positive aspects of sex are also undermined at every turn by the fact that partners can act in self-interest at the cost of their mate. We recognize this in the way we—and our gossip magazines—scorn cheating spouses, selfish lovers, parents who desert their families and partners who are unkind or violent. All of these expressions of self-interest threaten cooperative relationships and the societies within which partners live.

When I consider human sex and reproduction in the broader context of the animal world, I cannot help but conclude that we are a freakishly cooperative species. If animal mating systems were organized in a neat line representing the degree of sexual conflict, we would find some large seabirds, in which both parents work all breeding season to fledge a single chick, down at the end where conflict is minimized. At the other end there would be any number of insect species like the bed bugs in which males insert their equivalent of a penis straight through the female's body wall during mating, causing her physical trauma. Humans would be pretty close to the seabirds; human couples are hardworking cooperators who share a lot of the burden of raising a family. At least most couples are, most of the time.

Just as evolutionary biologists have recently realized that males and females often have quite different evolutionary interests, economics has had a similar epiphany. Economists modeling household decision-making long assumed that couples work to maximize their combined benefits. In the 1970s, economists and people working in various development agencies realized that men and women have different interests and that these interests often conflict. Adrienne Germain is now the inspirational head of the International Women's Health Coalition. Back in 1973 she was a 26-year-old sociologist working on family planning and development programs for the Ford Foundation, challenging economists to build models that incorporated the differences in interests between men and women within households. From her fieldwork in places like Peru, Germain argued that "every household has two decision makers, not one." The economists told her, "Look, Adrienne, if we were to add a second decision maker into our model, the equations would run off the paper. We can't do it." Germain's response was "Okay, you can't do it, so think of an alternative, because the reality is that household decisions are a negotiation."

Nowadays it is well understood that economic interests of men and women within households are often in conflict, and this new approach has led to important insights, particularly in addressing the role of inequalities in development. Nobel-winning economist Amartya Sen is one of those who argued that "cooperative conflicts" are a general feature of many interpersonal or intergroup relations, including family living. This interplay between cooperation and conflict is what makes relationships, particularly sexual relationships, so complex and so fascinating.

The link between evolutionary sexual conflict theory and economic cooperative conflict theory is more than a coincidence. In this chapter I explore the evolutionary basis of cooperation and conflict at the core of mother–father relationships and show how evolution creates the cooperative conflicts that economists model. A family is both an economic and an evolutionary unit. In most families, members benefit from one another's work, be it hunting, foraging, building, cooking, defending, earning, teaching, nursing or shopping. The joint benefits of the family's work can be measured in terms of economic wealth or evolutionary fitness, and the two currencies are intimately connected. For most of human history, people have worked hard and turned this hard work, and any wealth that came from it, into evolutionary descendants. When it comes to child-rearing, success is measured as the reproductive success of the children. It might only take a few minutes to conceive a child, but it takes almost half a lifetime of hard work to raise one.

Compared with almost every other animal that has ever lived, humans have evolved a very specialized lifestyle, which involves growing very large rather slowly. Our babies start out utterly helpless, unable even to walk, and they require close attention throughout a long infancy, childhood and adolescence. Parents, often helped by other relatives, provide children and adolescents with most of the necessities of life, teaching them where to live, what to eat, how to make a living and how to negotiate the infinitely complex ways of human society. It may take the proverbial village to raise a child, and an extended family or even one sprightly grandparent lightens the load, but in most societies the hard work falls foremost to the mother and sometimes to the father.

In the name of the father

Human families—where parents have a strong pair bond and share the work of raising a family—evolved from the single-mother family. We know this because in almost all mammals the young tend to hang out with their mothers. Sometimes they only stay around as long as the milk is flowing, but in mammals—like the elephants in chapter 3—where the young need to learn a lot in order to be successful adults, they often stay close to their mother for many years, learning from her everything they need to know.

Primates are unusual among mammals in that fathers often remain part of the social group to which the mother belongs, and in some species fathers are quite involved in the lives of their families. But in our closest living relatives, the chimpanzees, the family unit is a mother and her young. The mother conceives by mating promiscuously with most, if not all, males in the troop. So possible fathers are part of the troop in which the mother and child live and move, but no male can be certain that he is the father, and males don't help raise the young chimp.

Five million years ago, the ancestors we share with chimps were probably promiscuous and lived in smallish troops or bands. Females mated with many males around the time of conception, ensuring that any male in the band had at least a chance of being the father. But no male was confident enough of being the father of a baby that he so much as lifted a finger to help the mother out.

Since that time, our hominid ancestors gradually evolved to be less promiscuous than chimpanzees, yet much more promiscuous than people living in agricultural and industrialized societies today. Gradually, hominid men and women evolved the capacity to form pair bonds with one another; bonds that lasted a few years at best, and that were not exclusive. Most people probably enjoyed a varied sexual diet, yet they often shared the same dwelling and had most of their sex with one special mate. As these pair bonds evolved, those "special" men who had an especially high chance of being the father of a woman's children also evolved the urge to help out around her home—particularly by sharing food with her when she was pregnant and breastfeeding.

This blend of male interest and female promiscuity might seem like a bizarre and distant scenario but something like it flourishes in some places today. In some forest-dwelling cultures like the Aché of Paraguay and the Bari of Venezuela, women have a single husband but they also have sex with one or more other men when fertile, and the baby is considered the joint progeny of those men. In fact, according to Thomas Gregor, the Mehinaku men of the southern Amazon joke about paternity as a *wanaki*, an "all-male collective labor project."

In these societies that believe in partible paternity, children appear to benefit. Bari children with more than one father are better fed and about 80 percent survive childhood whereas only 64 percent of children with one father live to age 15. Historically, about one in five Aché children died violently at the hands of Aché adults, but children with several fathers were much less likely to die violently than orphans or those with only one father. When children survive better under a mating strategy, the mother's evolutionary fitness improves. So does the genetic father's fitness, but it is costly for a man to invest a lot of effort in children who are not his. As a result of the conflicting evolutionary interests of mothers and the various fathers, partible paternity is a fraught business that risks alienating the help or stoking the violent rage of at least one of the men involved. So it works best in societies where women do not depend too heavily or for too long on the labor of their husbands. These are often societies where marriages are brief and food is abundant.

Men have evolved an acute interest in the paternity of their children. Much of modern family life turns on the confidence men have that they are the biological father. Because a child gets exactly half its genes from the mother and the other half from the father, the evolutionary benefits of caring for children are almost identical for moms and dads. But for a dad this benefit is discounted by the fact that he is never quite certain that his sperm, and not the sperm of another man, fertilized the egg that grew into the baby in question. The more certain he is, the more likely he is to stick around and to invest in the child. The evolution of male parental care is very much the same story as the evolution of male certainty over paternity.

Much has been made of scientific reports that 10 to 30 percent of babies in industrialized societies are sired by a man other than the guy who thinks he is the dad. If this figure is true, then most married women must secretly be having enough promiscuous sex to make Samantha Jones from *Sex and the City* look like a cloistered nun. The idea of such rampant promiscuity titillates us because it undermines the foundations on which family life, and especially the investments fathers make in their children, are built. Accounts of rampant extra-pair paternity resonate in the public consciousness like an urban myth, but the reality is probably more mundane.

Many of the studies that report high levels of misattributed paternity are among groups where men already have grounds for suspicion: women whose husbands were off fighting wars, or men who have sought paternity tests because they have smelled a rat. Studies based on random samples, like screening for genetically inherited diseases, suggest that fewer than 1 percent of babies are sired by somebody other than dad. So, on average, dads get 99 percent the fitness return that moms get on any investments they make in a child's future. By mammalian standards those are good odds, and with the high certainty that fathers have in their paternity, caring for offspring makes good evolutionary sense. Most dads are much more than 99 percent sure that they are the father, yet a small number of fathers are less secure in their paternity—possibly because they have spent so little time at home, or because of the circumstances of their relationship with the mother. These dads who feel they have a good chance of not being the father are much more likely to abandon their family or to not put in the same kind of effort that average dads do.

New mothers, their relatives and even close friends work remarkably hard to reassure men of their paternity. The next time you see a friend or relative's baby for the first time, notice how suddenly you are gripped by the urge to point out how much the baby looks like the dad. Not that your reassurances are necessary in the modern world. For under $200, a suspicious dad can take mouth swabs from himself and a child, and send them to a lab for a screening test that will tell him with 99.99 percent certainty whether the child is his. Cheap and reliable paternity testing creates an explosive issue in those societies where it is available. By offering—for the first time in human history—accurate and inexpensive confirmation that a man is the father or that he has been cuckolded, these tests expose cavernous conflicts between the interests of fathers, mothers and children.

Should I stay or should I go?

Although the benefits of child-rearing are similar for mothers and fathers, two types of costs are very different indeed: the costs of *not* caring for the child are usually much higher for mothers than for fathers, but the reverse is true of the costs of committing to a largely monogamous relationship and staying to raise the family together. To become a mother, women really can't escape nine months of gestation, which takes an extortionate toll on the body, and ends with the perilous business of childbirth. A mother can choose to abandon the baby at birth or in infancy, and this sometimes happens, but by the time the baby has been born, the mother has a lot more of her future fitness to lose in giving it up than a father has in walking away. To begin with, it costs her at least a year to carry the baby to term and recover from the birth to the point where she can conceive again. One year wasted represents a significant chunk of her reproductive life span. So a father can recover from abandoning his mate much more rapidly than a mother could recover from abandoning the baby.

By staying and investing in raising a family, a man foregoes the chance to freely mate with other women. Men who stay and contribute to rearing the child pay an opportunity cost that is potentially much higher than that paid by the mother. A very attractive man could, in theory at least, impregnate hundreds of women in the time it takes any of those women to carry one baby to term. A father can decide at any time from ejaculation onward to walk out the door and not look back. Although he may have invested a lot of time and effort in courting and caring for the mother, it is possible that he could make up for his loss by having sex with many women within a few days.

A mother does not experience as big an opportunity cost because the absolute best that can happen, in evolutionary terms, is that she gets pregnant again but to a wealthier, kinder or genetically superior man. Even if she does that, she can still only have one baby in the coming year, and one every few years after that. In many, but certainly not all, circumstances the costs women would pay by abandoning the family far exceed the opportunity costs of staying. But the reverse can be true for men. As a result "deadbeat dads" are common enough even today to have a catchy name all of their own, but mothers who abandon their babies are headline news.

Given the very different costs and benefits of family life that men and women face, why do fathers bother joining the family at all? From a female's point of view, the potential benefits of having a male around the place might seem obvious: to help feed her and the child, take on some of the work she would otherwise do, and defend the family. But why would a male get involved in all this time- and energy-consuming domesticity when it was possible to keep on doing what other males had done for millions of years, leaving soon after sex in search of other females to mate with? Surely rampantly sowing of his proverbial oats would be a far more effective way to improve his reproductive success?

Men with pretensions to prolific sexual conquest are usually brought crashing down by the fact that the number of women who are available at any time is limited. Every baby has a mother and a father, and so the average number of children per man is the same as the average per woman. One point I return to often in this book is that although the average fitness of men and women is the same, the variation among men is usually greater than the variation among women. Some women do bear more children than others, but the costs of childbearing prevent most women from having more than a dozen or so babies. However, there are some kings, emperors and wealthy men who hit evolutionary paydirt by siring thousands of children. For every man who mates and sires children with hundreds of women, there must be several hundred men who never have the chance to mate. They may have died young—as young men often do in war, violence or showing off—or they may have lived but never had the chance to be a father.

Most men have a much greater chance of being an evolutionary loser than they have of ever holding a harem or reveling in the sex life of a rock star. The mere prospect of losing out completely is enough to make many a man happy to find a nice girl who will have sex with him regularly—even if it's only once a month, under the covers with the lights off. Our male ancestors probably also found such a deal too good to refuse. But how is a man to land such a catch? By his labor, of course. After all is it not a truth universally acknowledged that a single man in possession of a moderately good hunting ability must be in want of a wife?

For well over a century after Darwin turned his attention to human evolution in *The Descent of Man, and Selection in Relation to Sex*, the idea of "man the hunter" held a special place in our ideas of how human family life evolved. A mother needs a lot of food in order to make it through pregnancy and two to four years of breastfeeding, during which time she has very high energy and protein needs. So does a family of growing children. But mothers in late pregnancy, with breast-feeding infants or with dependent children cannot forage as much as they need to. How was a mother to procure such food? By seducing and thereby indenturing a tame hunter—a man.

Or so says the sex contract. The idea is that at the heart of the one-mom–one-dad family there is an unwritten deal: she will have regular sex with him and only him, bearing him children that he can confidently call his own; in exchange he will hunt and provide food, help out around the camp and protect her and the offspring from wild animals and something altogether more dangerous—other men. Monogamy under the sex contract is a convenient solution that guarantees regular men an exclusive sexual partner, and regular women the undivided industry and protection of a man.

As an idea, the sex contract has a lot in its favor. Sexual relationships, both long term and shorter term, often involve sex in exchange for material resources, the investment of time and energy in raising children, and protection. The sex contract, with its stereotype of man the provider, and woman the faithful and fecund mother, is a powerful idea that resonates with modern conservative notions of sex roles, the sexual division of labor and the nuclear family. In their book *Driven: How Human Nature Shapes Our Choices*, Paul R. Lawrence and Nitin Nohria extrapolate from the sex contract that early on in our evolutionary history "smart female hominids went to work on chimpanzee-like hominid males and—step by step, mate selection by mate selection—shaped them up into loving husbands and fathers with true family values."

This seems a little too hopeful and simplistic to me. The world is far more interesting and chaotic than this Wilma and Fred Flintstone version of the sex contract in which men have been literally domesticated by women. If men have been domesticated, then surely the job is only part done. And even when rearing children in monogamous nuclear family arrangements, the division of labor in human families is seldom as white-bread boring as portrayed in most versions of the sex contract.

Me hunter, you gatherer

This idea of "man the sexually contracted hunter" seriously overlooks the role of women as gatherers of most of the family's calories. Humans are not like hornbills—the spectacularly beaky Afro-Indian birds in which females spend weeks confined to the nest to incubate and raise the chicks, sealed in by a mud wall, with a small slit her only contact to the outside world. Hornbill males work relentlessly to provide all the food for the female and, once they hatch, the chicks. Unlike hornbills, the food women gather for the family is not only important, but it usually outweighs the food brought in by men.

In contemporary hunter-gatherer societies, and presumably also in our pre-agricultural ancestors, men and women bring different foods into the household to share. The stereotype is of man the hunter and woman the gatherer, and this pattern occurs often enough, but there are many other arrangements depending on the food types available in the environment. Mer Island on the far northern Great Barrier Reef between Papua New Guinea and the tip of mainland Australia is good fishing country, and the locals search for shellfish on the exposed reef, use nets and hand lines to catch fish from the beach, and fish from boats for larger reef and pelagic fishes. Fishing is by far the most important work toward subsistence among the Meriam, and both men and women participate in all three kinds of fishing.

Meriam men and women don't fish in identical ways, however. Rebecca Bleige Bird spent a year on Mer watching men and women fishing. She gathered detailed data on how long they spent, their success rates, and whether the catch was shared with other families. Women are more likely than men to collect shellfish, and to fish from the beach or on the reef in ways that target smaller reef fish—fish that are almost always there. Men are more likely than women to fish with spears or to target large pelagic fish like trevally or Spanish mackerel—fish that are sometimes abundant but are often absent. As a result of these differences, the weight of fish that women caught varied much less between fishing trips than did the fish caught by men. For men, fishing success was boom-or-bust with many poor days and a few bonanzas. The Meriam illustrate one universal in the division of labor between men and women: the food men and women bring in is never identical. They can overlap, but there are always some differences and those differences are quite stable within each culture. But how do such differences arise?

Maybe men and women divide their foraging effort in order to maximize their combined efficiency as a couple. If men specialize on some foods and women on others, the couple could bring in more calories in total than if each had to gather all the foods on which the family depends. According to some economic models, even cooperating individuals of identical ability can slip into specialized roles as each individual becomes highly efficient at some tasks. Small differences can consistently influence which roles one partner performs, leading to stable but flexible roles. Among the Meriam, men's slightly greater size and strength, and the reduced mobility and speed of women who are pregnant, breastfeeding or have to keep an eye on small children, could make it best for men to do the spear fishing and trolling for large pelagic fish. Similar differences might make women in hunter-gatherer societies more effective gatherers of plant-based food than men.

But individuals working for selfish gain inevitably find ways to undermine this kind of optimistic scenario. Men and women have different evolutionary interests, making some types of foraging effort more profitable to one sex than the other. Division of labor can depend more dramatically on male–female differences in reproductive strategy than on maximizing household productivity. Women have more at risk from a period of starvation than men do because children and unborn fetuses are at greatest risk of death, and every mother's reproductive fitness is more tightly tied to the survival and well-being of her existing children than is the father's fitness. Most families, especially in hunter-gatherer societies, depend on mom bringing home most of the daily calories. This aversion to risk forces mothers to focus on the most reliable, if not the most spectacular sources of food, just as Meriam women target the more reliable fish and shellfish species.

Whereas most of the fish caught by Meriam women go directly to feeding their families, the big spectacular fish that men occasionally catch are more likely to be part of the tradition of elaborate feasts, which, according to Bird, "involve dancing, distribution and display." This difference is typical of many hunting societies in which the food that women gather or catch directly feeds their families whereas men divide the meat from any large animal they have killed among all the households. The larger the prey that men chase, the more likely it is to be shared around. Bird and Kristen Hawkes have argued that hunters who do this are acting more selfishly than it appears, using the division of spoils from a kill as a way of showing off to women other than their wives, and earning extra matings.

By hunting or fishing for prey that are small, abundant and easy to catch, men could bring valuable calories home to their family on most days. Instead, men often hunt the largest and most elusive animals: large pelagic fish among the Meriam, bison in precolonial North America, armadillos and peccaries in South America, kudu, zebra and eland in Africa, and possibly even mammoths in ice-age Europe. By chasing such big prey, men actually bring in far fewer calories than they would have done by targeting smaller, easier to catch and less dangerous animals. Instead of aiming for hunting efficiency, men target large animals because by killing such a beast and sharing the meat they burnish the legend of their hunting prowess, and this leads to better mating opportunities in the future. An extra mating can far outweigh, in evolutionary terms, the mundane fitness advantage of providing a few thousand calories to one's own wife and children. Much of what we know about life in prehistoric hunting societies comes from ancient cave paintings of hunts, and there is precious little prehistoric art of gathering. It strikes me that many cave paintings may have been attempts to keep fresh the memories of the most epic hunts long after the meat had disappeared.

Still unequal after all these years

In bird species like albatrosses and eagles where parents have to work extremely hard just to fledge one chick, a female often pairs up with the same male every year, and few if any of her chicks are sired by another male. Something similar happens in human societies. A family's ecological and economic circumstances—how difficult it is to eke a living from the land or from the economy—influence how men and women divide their labor, and how long marriages tend to last. There are much bigger differences among societies and among families within those societies in how hard men work than in how hard women work. Women—at least mothers—tend to work as hard as they can. When living is easy, men often do as little as they can toward the household. They enjoy more leisure time and spend it doing manly things like hanging out, drinking, smoking, competing in sports, fighting and raiding other villages for new brides.

For hunter-gatherers and horticulturalists, living is much easier in the tropics where there is plenty of sunshine, enough rain and not much seasonality. Plants grow spectacularly, supporting plenty of animals, so people find it relatively easy to eat a balanced diet without working too hard. There is enough food in the environment for women to bring in most of the family's needs, and additional food brought in by men is not all that valuable to the woman and the family. The Mehinaku typify this pattern; they inhabit productive forest in the southern Amazon where they catch fish, gather fruit and clear forest patches to grow manioc and maize. Women work about eight hours a day at gardening, processing food, making household items and collecting wood and water. Men work less than half that, on average, mostly fishing and gardening. They then spend a lot of time hanging out in the village "men's house" with other men as well as in daily wrestling contests.

Compared with cushy tropical lifestyles where men are free to pursue leisure and warfare, men in more seasonal places where there are fewer palatable vegetable foods to gather work much harder. In these places families work harder to make a living and men's contributions to the family pot are relatively more valuable to women. There is also often a clearer division of labor. The Inuit of the Canadian Arctic embody the division of male hunters and females working in the home. Men hunt for seals, walrus, whales and caribou to provide almost all the food and raw material for tools and clothes. Women butcher the prey, prepare the food, render oil for heating and lighting, and chew and sew skins. Husbands and wives are tightly interdependent and neither can do their work without the hard work of their spouse. It is only as a team that Inuit families can survive in the most extreme environment people have ever inhabited.

The harder that men work, the longer marriages tend to last. In tropical environments where women don't rely heavily on the calories that men provide, marriages tend to last a few years at best—long enough for women to make it through the critical period of pregnancy and breastfeeding during which the food men provide improves the baby's chances of survival. Among the Aché, who live in a similarly productive Amazon rainforest habitat to the Mehinaku, women go through more than ten different marriages—each lasting a few years—before menopause. In tougher environments where women depend more on their husbands, men not only work harder, but those that do, stay married for longer. Divorce is much rarer among the Inuit than in tropical societies like the Aché.

Not only do men work only as hard as they need to at providing food, but men are adept at providing as little parental care as they can get away with. The anthropologist Sarah Blaffer Hrdy has done as much as any other scientist to help us understand human family life in the context of evolution. In *Mothers and Others: The Evolutionary Origins of Mutual Understanding*, Hrdy generalizes that the investments fathers make in their children—both by provisioning the family and by interacting with children directly—are way more variable and depend very much more on circumstance than do the contributions made by mothers. For example, when the mother is among her own relatives, men often ease back on the time and effort they spend caring for children—leaving their wives' mothers and sisters to pick up the slack. When a couple lives away from either set of relatives, men contribute more on average around the home and especially to caring for the children.

The same is true in modern industrialized societies where sociologists and economists have long observed that women work longer hours toward the joint household, on average, than men do. But what happened to the rise of the working mom and the twenty-first century superdad? For the last 40 years at least, sociologists have debated whether marriages are becoming more equal or not. Many modern men are certainly taking on more child-rearing responsibility than their fathers did. The number of women in full-time work continues to increase steadily, and although modern working women continue to be underrepresented at all senior levels, their representation also continues to improve. As a result some authors argue that modern societies are on a trajectory toward real equality within households. The problem is that in the 1970s, those authors predicted we would have genuine equality within marriages by the year 2000. But a decade later we remain a long way from this nirvana, even in countries with the greatest gender equality.

The increase in men's contributions to housework and child rearing does not offset the gains in women's paid employment. As mothers enter the paid workforce, many simply add paid work to their existing childcare and household responsibilities. Some fathers do take on more cooking, cleaning and caring, and some of them cut back on the number of hours or days they work in their jobs, genuinely freeing up their partners to work. But if you look at a country like Australia as a whole, and add up the hours spent in both domestic and salaried work, women outwork men. In Australian families where both men and women are in full-time employment, women do 71 minutes more housework every day than men. In fact, these men only do 12 minutes more domestic work each day than working men whose partners are not in paid employment. And little has improved for stay-at-home moms whose unpaid work remains undervalued by their partners.

Men and women also spend their hard-earned money differently. In wealthy societies men spend more money on leisure, entertainment and luxuries than women do. A man is more likely to maintain his social life, retain the same friends, and enjoy playing and watching sport after becoming a parent than a woman is. Even when they work 14-hour days and make barrels of money, it seems men's work is not always directed solely at improving the lot of the family. Fathers still spend money on signals of status and wealth like luxury motor vehicles, and on recreational goods like boats and fishing gear, even though they have already wooed and won a mate.

This difference in spending patterns is embarrassingly similar to the way hunter-gatherer women work long and hard to keep the wolf from the door by bringing in unspectacular but dependable foods, while men go off on ambitious yet seldom successful hunts to show off their hunting prowess and share around. But not only hunter-gatherers and rich Western men behave this way. Adrienne Germain notes that in the poorest developing nations, where subsistence farmers scratch a living from the earth:

> when a woman has income at the family level, she does allocate most of it for her children and the basic survival of the household, whereas the expenditure surveys with men clearly document the leisure time, the cigarettes, the alcohol, the . . . unnecessary clothes, like western-style shirts.

It seems that every joke or hackneyed stereotype about lazy, self-serving men might be true. Of course a great many men are wonderful fathers, exceptionally hard workers and tremendous partners, but it remains stubbornly the case that most fathers do less, often much less, than most mothers do toward the well-being and success of the family. It seems the answer to my question about what men get out of confining (most of) their attention to a single woman is that they get quite a lot. Men who settle down with a woman and put in at least some effort can be pretty confident that they are indeed the father of her children. They benefit not only from largely exclusive access to her womb, but also from the work she invests in caring for, feeding and teaching their children.

But what does the female get out of it if most men don't put in an equal effort? The answer depends on a slight shift in perspective. What matters is not whether the work is shared equally, but whether each participant fares better from being in the relationship than they would have done from being without. Men and women have evolved strong pair bonds and ways of working together to raise families in order to benefit from one another's contribution. Where the benefits of staying and working outweigh the costs for both parents, they should stay together. Because these benefits and costs differ for the two partners, moms and dads will seldom contribute equally. So the relationship can have a different value for the two partners, and other circumstances such as the health, age and independent material wealth of the two partners can also contribute to differences in how they value the relationship.

The balance of power and home economics

The male–female division of labor, especially the highly visible labor of bringing food and wages into the home, also influences the power that men and women wield in the home and within society. In those hunter-gatherer societies where both men and women have to work hard, women tend to be valued and regarded within their societies much more highly than they are in societies where men do less than women. Societies in which men hunt big game are often societies in which women don't depend on their husband's hunting as much. In these societies, a man's shared hunting spoils make him more likely to enjoy a little romance on the side, and they elevate his political clout within the village. In societies like these, women tend to wield less power than they do where men work alongside them gathering vegetable foods, catching fish or small game.

The earliest moves toward agriculture probably improved the lives of most women. In some traditional societies, particularly in the Amazon and much of sub-Saharan Africa, families garden patches of cleared forest or bush, growing starchy staples like manioc, taro, yams and maize as well as fruit trees, and supplementing this diet by hunting, fishing and gathering foods. Agriculture got started via horticulture and gardening, and because women tend more often to be the gatherers, it was probably women who first worked out how to cultivate the plants that eventually evolved into crops. In these horticultural societies women still work more hours a day than men, but hard work on the part of women tends to be rewarded by hard work on the part of men—especially when women are allowed to divorce a lazy husband. The ability to escape a bad marriage together with the fact that a woman's contribution to the daily diet is highly visible gives women power in the home and in society that is not all that much lower than the power men wield.

When women are responsible for the bulk of food production as they are in gardening and simple agricultural societies, women also often live close to their female kin. Primate mothers fare better in the cooperative conflicts of reproduction when they live close to their own mothers and other kin than when they move a long distance into troops or harems where they have no kin. The same tends to be true in humans. Having kin around makes a woman's life easier because she has allies with an evolutionary interest in helping her raise her family successfully. Her own mother is especially important, helping to care for young children and bring in extra food. When women consistently live close to their own parents (called matrilocality), they generally tend to enjoy much greater autonomy, more equal power with their husbands, greater influence in the affairs of the society, and lower levels of violence and spousal abuse compared with societies where women live close to their husband's family (patrilocality).

Women in patrilocal societies are less likely to venture outside the home, or hold paid employment or positions of influence. In northern India, for example, brides tend to move to their husband's natal village, which is often far from their own family and friends. They live close to their husband's parents and brothers, and many are compelled not only to look after their own children but to cook and attend to the needs of their in-laws. Remember that the man's parents and siblings have evolutionary interests that are closely aligned to his interests, and they will often interfere in the couple's affairs in ways that suit their own and their son's interests. It is little wonder that few things strike as much terror into the heart of a young wife as the imminent materialization of her mother-in-law.

The Iroquois people, who inhabited the area surrounding Lake Ontario at the time Europeans were settling North America, illustrate how matrilocality and a prominent role in food production elevate the status and power of individual women. Iroquois women stayed in the same settlement throughout their lives, sharing a longhouse with other female relatives and their families, whereas men moved among settlements to marry. Men covered great distances to hunt, fish, and fight or conduct diplomacy with other clans and other tribes. Women worked the plots of domesticated maize, beans and squash, which made up most of the Iroquois diet, and mothers taught their daughters how to farm and use agricultural tools. Families in which women stayed close to their mothers and sisters would have made superior farmers because of the economic advantages that came from collaboration among women in agriculture, and these matrilocal clans would have outcompeted and displaced those clans in which sons stayed close to their brothers and their wives made a less cohesive agricultural workforce.

Although the work that men and women did was different, Iroquois society is often held up as a paragon of sexual equality. Men held the prestigious positions of ceremonial and political power, but they were beholden to the women of their clans who decided which men would represent them and could remove them from their positions. Men acknowledged and appreciated women's work as providers, and women owned property and land and passed on their property to their daughters. It was relatively easy for a woman to get rid of a lazy or difficult husband, forcing him to leave the longhouse and take his possessions with him. European settlers also noted that rape and physical abuse of women was practically nonexistent, even when a clan captured the women of another clan in warfare.

Even though early agriculture probably elevated women, as farming became more intensive the lot of women worsened steadily. The sedentary farming lifestyle freed women from having to carry their babies and therefore having only one dependent child at a time. Extra hands were needed on the farm, and the families that bred fastest built the largest workforces, allowing more land to be farmed and more wealth to be made. The extra food also fed the rapidly growing family. As we saw in chapter 4, where natural selection previously favored modest reproductive rates with long gaps between babies, suddenly the fittest women were those who pumped out babies as fast as possible. As many feminist authors have pointed out, being a homebound baby-making machine is hardly empowering.

With the domestication of large animals and the introduction of the plow, men suddenly took on a much more substantial role in the production of food than previously. The art of ancient Egypt and the Levant shows men doing the plowing and herding. Conventional thought holds that men, a little taller and stronger than women on average, were more effective at these tasks. But crop surpluses and livestock numbers became history's first forms of significant wealth—and wealthy males, like conspicuously good hunters, probably got many more matings. Perhaps men commandeered agriculture precisely because the incentive arose; those men who became wealthy enjoyed enormous fitness benefits by having many wives and many children. Those men also probably did better if they obtained the food, and left it to their homebound wives to bear and raise the babies.

It is hard to overstate how profoundly the sudden ability to store food surpluses, raise livestock and turn both into wealth changed human society. Animals, grain and money can be stolen, and people can be displaced from their productive lands, so people needed to defend their land and their stuff. Men had the most to lose if their land or wealth was stolen, and men also had the most to gain from stealing. Defending resources requires allies, and a man's best allies are his brothers and sons with whom he shares a genetic interest. So families in which men stayed on at the family farm, helping to expand and defend it, made and held on to considerably more wealth than families where daughters stayed home and brought their new husbands to live. These developments also made it more advantageous to pass wealth on to sons rather than daughters, and so inheritance customs often came to favor sons.

The accumulation of wealth also gave rise to the formation of wealthy and powerful elites: chiefs, kings, military leaders, and priests. These lucky few made and enforced the rules, both celestial and terrestrial. Because the circumstances that begot agricultural wealth and then power were also circumstances that consigned most women to relentless childbearing and child-rearing, the most successful agricultural societies also often tended to be ones where most women held little power within society. As a result of these simple forces that allowed men to seize the power to make laws and interpret the will of the gods, laws came to favor the interests not only of men, but of wealthy men. Women often became possessions of their husbands, and poor men became the rank and file of ever escalating armies. Wealth also made it possible for some wealthy men to have many wives, and to keep harems of fertile women for their own sexual gratification and reproductive advantage. Thus did the success of agriculture give rise to the patriarchy.

Moving forward

The Industrial Revolution came to Europe just in time. In the late eighteenth century, arable land in Britain and Europe was more or less saturated, population growth was outstripping food supply and Thomas Malthus was foreshadowing the imminent famine, plague and death on a scale never before seen. As we saw in chapter 4, the nineteenth century saw a shift from quantity to quality as a parenting strategy. Whereas, before, having more children meant more hands to labor on the farm, now they merely meant more mouths to feed. Suddenly having fewer children and educating them as well as possible became the main route to upward social mobility. This kind of intensive investment in fewer children required hard work from both men and women, tipping the cooperative conflict somewhat closer to the interests of women than it had been throughout millennia of agriculture. Gradually women's roles changed from high-output baby-making machines to earning wages and teaching children, at least among the working class.

It is no coincidence that in the late nineteenth century women in many societies made the greatest and most widespread political gains that they had ever collectively made. The Industrial Revolution wrought tectonic economic and social upheaval on society. Much of this change was negative, but the valuable labor of working-class women and the importance of investing in children raised the visibility and the value of the work that women did within and outside of the home. At exactly the same time, the first-wave feminists were claiming for all women the same rights that men had long held: to own property, participate freely in the economy, vote and run for representative office.

Biology and feminism have seldom made comfortable bedfellows, if you will pardon the metaphor, but it seems clear to me that the differences in biological costs and benefits of reproduction are right at the center of the issues that concern feminism. Insights into the conflicting agendas within families have redefined the economics of development such that it is now widely understood that the economic empowerment of women is critical in combating poverty, child mortality, HIV and population growth. And the technological and legislative developments that have improved the lot of women in industrialized societies have done so by helping to cut the biological costs of being a mother. Fifty years ago, the pill let women take unprecedented control of their fertility. Access to safe and affordable abortion allows women to avoid becoming mothers at a time or in circumstances that would be detrimental to them. Paid maternity leave allows women to spend time out of the workforce being mothers while continuing to provide for their families, and to return to the meaningful and rewarding jobs that they held before they gave birth. No-fault divorce allows women and men to escape from marriages where the balance between conflict and cooperation is wrong.

Many social conservatives lament the effects that these developments have had on modern family life. Divorce rates rose steadily through the twentieth century in industrialized nations, and marriage rates continue to plummet: these are both trends that the churches and conservatives, who favor a definitively masculine view, bewail. I find it impossible to swallow the idea that we are worse off for these changes. The proportion of people who are married is no barometer of the well-being of a modern society. In Australia we recently elected our first woman prime minister, Julia Gillard, and I was pleasantly surprised at the maturity with which our media treated the fact that she is unmarried and lives with her partner. Married or unmarried seems increasingly like an irrelevance to those of us not tethered to the antiquated religious establishment. In fact, old ways of defining relationships and sex roles seem more likely to get in the way than to help women and men as they renegotiate their partnerships and parental roles in the unprecedented economies and environments of the twenty-first century.

7

Love is a battlefield

I think serial monogamy says it all.

—Tracey Ullman, 1989

What is the natural human mating system? Are we monogamous, polygynous, polyandrous or even promiscuous beings? People are all of these things, some of the time, but there is no one-size-fits-all human mating system. Mating systems also change societies. I consider the frightening social consequences that occur when powerful and wealthy men are unfettered in their ability to marry or otherwise monopolize the reproduction of several women.

JACOB ZUMA, PRESIDENT OF SOUTH Africa, comes from the humblest beginnings. Zuma has only five years of formal education, but a lifetime of experience hard-won in the trade union movement, the anti-apartheid struggle, and as a prisoner on Robben Island with Nelson Mandela. Whereas his predecessor, Thabo Mbeki, always radiated cool statesman-like dignity in finely tailored suits, Zuma can still sometimes be seen in traditional Zulu leopard-skin outfits belting out the anti-apartheid rallying song "Lethu Mshini Wami" (translation: Bring Me My Machine Gun). Zuma is a controversial and enigmatic figure; he rose to the presidency as a populist, stoking the ambitions of those who feel they are little better off now than when apartheid ended, but in office he is finding out just how difficult it is to govern. Since 2004 Zuma has been tainted by a massive corruption scandal over an arms deal, yet he somehow survived and outmaneuvered Mbeki to oust him as president.

Zuma governs a country that is as rife with contradictions as he is. The dismantling of apartheid, and Nelson Mandela's 1994 election as president, were nothing short of miraculous, yet South Africa remains a country of stubbornly persistent inequality and endemic violence. It had the modern infrastructure and organizational know-how to flawlessly host the 2010 FIFA World Cup, yet traditional and often anachronistic customs of dozens of ethnic groups are sacrosanct. It has some of the most progressive gender-equity legislation on the planet and a high representation of women in parliament, yet sexism is rampant, the rape statistics terrifying and more than 5 million people are infected with HIV. Zuma put South Africa's HIV/AIDS policy on the right track after years of denial and dithering by Mbeki, yet in his private life Zuma embodies ignorance. He was acquitted of the 2005 rape of an acquaintance who was known to be HIV positive, with the judge accepting his defense that the sex was consensual. Zuma admitted that rather than wearing a condom, he showered after sex in order to reduce the chance of HIV infection.

In addition to his much-publicized extramarital promiscuity, Jacob Zuma has been married five times. But unlike celebrity serial monogamists like Elizabeth Taylor, whose many consecutive marriages titillated Western gossip mags, Zuma is a polygynist with three current wives. His second wife, Nkosazana Dlamini-Zuma, herself a senior government minister, divorced him in 1998. Zuma's third wife committed suicide in 2000. Yet the president does not appear to be stopping his marrying ways. He has paid *lobola*, the traditional bride-price that signifies engagement, to the families of two more women. By his wives, fiancées and at least four other women, Zuma has allegedly sired at least 22 children.

In evolutionary terms, polygyny and extramarital vigor are delivering Zuma high fitness.

But his polygyny is controversial. When Zuma made his first state visit to the United Kingdom in 2010, soon after his fifth marriage, most of the press coverage centered on his polygyny. But why should the Brits or anybody else care about the exotic domestic arrangements of a foreign head of state? Are they, as Zuma's supporters allege, just petty cultural imperialists intent on imposing their own bourgeois values on other cultures?

Love in the natural way

Mating arrangements and marriage customs vary enormously among societies, from strict lifelong monogamy, to shorter monogamous unions lasting a handful of years, to polygyny in which some men marry two or more women, to polyandrous marriages in which a woman takes two or more men—often brothers—as husbands. But what is the "natural" human mating system?

Some anthropologists and sex researchers point at our capacity for deep and profound pair bonds, and claim that humans are naturally monogamous. Helen Fisher, one of the most influential popular writers on human sex, asks, "Is monogamy natural?" Her answer is, "Yes . . . among human beings polygyny and polyandry seem to be optional opportunistic exceptions; monogamy is the rule." This message resonates with conservatives and fundamentalists, many of whom would have problems saying the word "evolution" without choking on their own bile, yet who cheerily assert that lifelong monogamy is the natural state of affairs. Those who stray from the path of one-woman–one-man righteousness are not only sinners, according to the ancient gospels, but they also defile the order of nature.

But the evidence is just as strong that we are naturally polygynous and that men can love and bond with many wives, whereas women tend to bond with just one husband. This position is understandably popular with polygynous men and those men who aspire to emulate them. More than four out of every five societies allow one man to be married to more than one woman at a time. It might appear then that polygyny is pretty much the norm. Yet, only a minority of men in polygynous societies have more than one wife. Many small polygynous tribes of a few hundred souls inflate the number of societies that permit polygyny. Indonesia, with 250 million people, is the most populous nation where polygynous marriage is generally allowed, but most people alive today live in larger societies that do not legally permit polygyny.

To add to the confusion, good evidence also indicates that people are naturally promiscuous. In their refreshing recent book, *Sex at Dawn*, psychologist Christopher Ryan and psychiatrist Cacilda Jethá point out that while we fall in love and form pair bonds, we also relish plenty of sexual variety. Marriages in many of the remaining hunter-gatherer and horticultural societies tend to last only a few years. Many married men and women in modern societies entangle themselves in several sexual relationships at once, and Ryan and Jethá argue that our ancestors did the same, at least until the very recent advent of agriculture allowed men to accumulate property and wealth. Not only do humans sometimes love and mate with many partners in a short time, we also have a fine-tuned interest in "who's doing whom," and a murderously acute capacity to detect infidelity.

In my opinion there really is no such thing as a single "natural mating system" for humans. Mating systems such as monogamy, polygyny and polyandry are mere pigeonholes that help us organize the mind-boggling variability we see in the world. Evolution shapes the repertoires of behaviors that individual men and women are capable of, and especially how men and women use and adjust those behaviors to suit their circumstances. As environmental circumstances like the abundance of food or the ratio of men to women change, so individuals will adjust their behavior. As a result, the aggregated effects of modest shifts by individuals can effect quite big changes in broad patterns of behavior.

Even within a single "mating system," there is always a lot of variation among individuals. For example, in a so-called "polygynous society," usually fewer than 10 percent of men have two or more wives. Most men who are married are in monogamous marriages, and there are many men who are not married at all. Likewise there are unmarried women, monogamously married women and women who are one of many wives married to polygynists. Every one of these individuals is doing what they can to thrive and, usually, to reproduce. And our ancestors have negotiated an even wider range of roles in every generation since sex evolved.

The institutions of marriage by which human societies regulate responsibility for children and property ownership do shape our mating patterns, but we should not be blinkered by these conventions into believing that they define our mating system. Even in Western societies that allow only monogamous marriage, most people have more than one sexual partner throughout their lifetime. In fact most people have, at some time during their lives, more than one sexual partner at one time. Men and women in monogamous long-term relationships or marriages seek out or fall into affairs and sexual liaisons: some that are short-term and some lasting many years. These affairs are so common that strict sexual fidelity to one partner is probably by far the exception rather than the norm.

Even a simple category, like polygyny, can be richly complex; polygynists come in many shapes and forms. Jacob Zuma, who is married to many wives at the same time, and the late Nigerian musician Fela Kuti, who married 27 women in a single ceremony, can both unambiguously be considered polygynists. But marriage is irrelevant in evolutionary terms. From an evolutionary point of view, Tiger Woods, who married Elin Nordegren yet had affairs with many other women, is also somewhat polygynous. And Rod Stewart, who, in succession, either married or had long-term relationships with the models Dee Harrington, Britt Eckland, Alana Hamilton, Kelly Emberg, Rachel Hunter, and, most recently, Penny Lancaster-Stewart, is himself a model—for the special kind of polygyny we euphemistically call "serial monogamy." There are differences, but only in degree. In modern Western societies where people routinely enter and leave many long-term sexual relationships, nearly all men are polygynists—even if they are never unfaithful.

The same applies to women. In only a handful of societies can women marry more than one man at once. The best-studied polyandrous society nestles high in central Tibet, where a woman marries a group of brothers who together contribute to the household and act as father to the children. This especially rare strategy is practiced only by landowning farming families in steep valleys where there is no more land to occupy. With brothers unable to divide the farm without impoverishing all of them, and with few prospects for one brother to successfully establish himself elsewhere and take a wife, brothers enter a polyandrous marriage to one woman. Yet women who enter serially monogamous marriages, like the nine-times married Elizabeth Taylor, are polyandrous even if they appear monogamous at any given time. And having affairs outside of marriage is another way in which women often commit polyandry—at least from an evolutionary point of view.

The natural history of human mating and marriage is a topic at the confluence of four themes of this book. First, observations of how evolution has shaped societies are simply snapshots of a process that acts on individuals and their genes. If we want to understand what is natural, we need to look at the interests of the individuals involved—not just the two (or more) people who are getting it on, but the partners and families whose futures are intertwined.

Second, I am firmly among those who argue that biology alone cannot prescribe what is right, what is wrong, and how society should function. But I am equally firmly of the opinion that an understanding of evolution is indispensable in resolving how society came to be the way that it is, and how we arrive at our *ideas* of right and wrong. Thinking typologically about humans as a species, or even of men and women as if they each were the uniform inhabitants of different planets, can only take us so far. Humans are not only capable of, but quite thoroughly practiced in, just about every kind of sexual partnership, liaison, deception or intrigue that can be imagined. This variety is a product of evolution, but not in the stereotyped "hardwired" way that moralists use as they ineptly fumble for a naturalistic imprimatur to their own favored way of living. To claim that as a species we conform naturally to one or another pattern—and not the other patterns that we are capable of—is either denial or deception.

The third theme is that the evolutionary interests of individuals—including the interests of an individual woman and the individual man she mates with—are often in furious conflict. And the last theme is that men and women have evolved capacities to make out as best they can under the environmental circumstances into which they are born. I have already discussed these two themes in chapter 6 where I explored how pair bonds and the cooperative conflicts between mates evolved from the promiscuous mating system of our early hominid ancestors. And it is this thread that I must now pick up.

Evolution of polygyny

According to Ryan and Jethá, polygyny is a newish phenomenon, arising only when agriculture allowed men to control most of the resources in a society, and "suddenly women lived in a world where they had to barter their reproductive capacity for access to the resources and protection they needed to survive." I disagree. Although agriculture and the resulting shift to unequal male-controlled wealth gave men unprecedented power to constrain and control women's sexuality and reproduction, this is simply an embellishment of a story that is as old as humanity.

The promiscuous bedrock of our pre-agricultural history is still visible in contemporary horticultural and hunter-gatherer societies, especially those in tropical ecologies where living is relatively easy. In many of these societies, like the Aché and the Mehinaku, partnerships and marriages typically don't last long, dissolve easily, and there is plenty of extramarital hanky-panky. In deserts, or at high latitudes like the Inuit and Athabascans of the Alaskan Arctic, couples depend heavily on one another's labor, marriage lasts a long time, and both partners work extremely hard just to get by.

One might expect that polygyny would be concentrated in the foraging societies where men work hardest; after all the hardest and most resourceful working men—the best husbands—should be in high demand. Instead, polygyny is most common in cultures where men are laziest of all. In marginal high-latitude and desert environments where husband and wife form a tightly co-dependent economic team, monogamy is the norm. In the tropics, where men don't contribute nearly as much as women to feeding the family, polygyny is common. Men, with spare time on their hands, turn their attention to competition with other men, including warfare with other villages and establishing positions of high status and dominance within their village. The polygynists are usually the individuals who have risen to dominant positions like chief, shaman or witch doctor.

In many societies women are coerced into marriages, both monogamous and especially polygynous unions. Among the nomadic foraging Aborigines of Arnhem Land, in northern Australia, polygyny was common until missionaries arrived in the mid-twentieth century. In one settlement, studied by Jim Chisholm and Victoria Burbank, men who married monogamously were, on average, around 7 years older than their wives, but polygynists were around 17 years older than their wives. In this society, old men of high status oversee young men's initiation into society, control who can practice in the religious cults, and confer on mothers the authority to choose husbands for their daughters. Unsurprisingly, these are the men who end up marrying the most women.

The Arnhem Land Aborigines studied by Chisholm and Burbank typify another common pattern in societies that allow polygyny. Men jockey aggressively for status and power, which they then wield to obtain as many women as possible. The very possibility of having many wives drives men to compete violently with other men and to forcefully coerce women. According to Chisholm and Burbank: "there are abundant recorded instances of Aboriginal men resorting to very literal coercion—violence—in order to obtain a wife, to prevent a wife from leaving them, or to force her to end an affair."

Long before agriculture arose, men and women evolved the capacity to respond to inequalities in status, wealth and power. The inequalities were always there, but they became much more pronounced with the dramatic inequalities that agriculture brought forth. As landowners accumulated wealth and found ways of passing on that wealth and their land to their sons, so wealthy men gradually installed themselves as landlords, leaders, lawmakers, slave owners, peacekeepers, taxation authorities and the terrestrial interpreters of divine will. The bigger the inequality in a society, the more tightly the elite men controlled the power. And they wielded this power to lock their wives into ever more exclusive matrimonial arrangements, and to add to the number of their wives and concubines.

Throughout history it is never the poor peasant laborers that had the most wives, but rather the wealthiest men. Biblical accounts of King Solomon tell us he had not only great wealth but also 700 wives and 300 concubines. More recently, King Sobhuza II, who ruled the tiny and impoverished African kingdom of Swaziland between 1921 and 1982, and was its wealthiest citizen, married 70 wives who bore at least 210 children. People living in different places and at different times respond in predictably similar ways when inequalities in wealth and power arise, suggesting that our responses to inequality are ancient evolved traits. The ways men and women responded to the unprecedented inequality that agriculture opened up are simply embellishments of earlier adaptations to profit from the smaller inequalities that occurred among our foraging ancestors and their contemporaries.

Make war, not love

Herbert Spencer was one of those Victorian gentlemen who euphemistically saw reproduction as a means to perpetuate the species. As long ago as 1876, Spencer made the very useful observation that polygynous marriage is more common in societies where there are high male casualties in war, leaving many women per surviving man. But he then leaped to the conclusion that polygyny is an adaptation to maximize the reproductive rate of those societies and replace the casualties of war. With many women per man, Spencer argued, societies that encouraged polygyny would grow faster because each of those women would be able to reproduce. He failed to recognize that it is not society that imposes polygyny from the top down to suit its amorphous ends, but polygyny that emerges from individual men and women acting in evolutionary self-interest. Polygyny, perhaps more than any other practice, exposes the flaws in old Victorian-era thinking that fails to see the rampant conflict in the midst of sexual reproduction.

Of all the individuals touched by polygyny, the interests of the polygynous men are easiest to decipher. Unsurprisingly, men who marry many women win big in evolutionary terms. By marrying a second wife, a man can double the number of legitimate children he fathers. Extra wives also add to the domestic workforce that services his needs. Marrying several wives spectacularly breaks the shackles that monogamy places on a man's reproductive success. Men have evolved to strive for the chance to marry many women whenever it presents itself; these are the men who make up a large proportion of our ancestors.

Unmarried men are big losers when societies allow polygynous marriage. For every man with two wives there is one man who never marries; for every man with three wives there are two men who never marry, and so on. Some of these evolutionary losers live long lives without marrying, but many men who never marry simply don't live long enough to do so. As Spencer noted, there is a direct relationship between polygyny and violence. The more widespread polygynous marriage is within a society, the more likely it is that the society will go to war with neighboring groups or tribes, and the higher the levels of assault and homicide within the society.

Violence is both a cause and a consequence of polygyny. It is a cause because warrior societies lose a lot of young men in conflict, leaving many more surviving women than men. As a result, some women cannot find their very own man, and are forced to choose between not marrying at all or becoming a second or third wife. When men are scarce, even the least desirable among them can often find at least one wife, and many can find a second or even a third.

Warfare is also a disturbingly direct way of accumulating wives. Much ancient warfare involved the raiding of other villages to capture fertile women as brides. The most warlike groups of men with many strong warriors can defend their own villages and women, raid other villages and capture brides. The fiercest groups quickly become the most polygynous, and the fiercest warriors take the most wives.

In men the drive for conspicuous status, wealth and dominance over other men co-evolved with the male appetite for polygynous marriage and for plenty of extramarital sex. Men bear signs of a past in which selection favored those prepared to risk everything to get ahead. If there were a contest for the biggest evolutionary winner in history, I would bet a month's pay on Genghis Khan. According to the warrior who founded and ruled (from 1206 to 1227) the largest contiguous empire in human history, "the greatest happiness is to scatter your enemy, drive him before you, to see his cities reduced to ashes, to see those who love him shrouded in tears, and to gather into your bosom his wives and daughters." Modern DNA evidence shows that the bit about gathering of wives and daughters of the vanquished to his bosom was indeed a big part of Khan's motivation for conquest. Khan carried a distinctive mutation in his Y chromosome, the small piece of genetic material passed only from father to son. Today, one of every eight men in Asia and one in every 200 men worldwide—a total of around 16 million men—bear this mutation and can trace their descent through the male line directly back to Genghis Khan.

Khan and his male descendants conquered an area from the Sea of Japan to the Mediterranean, never wasting an opportunity to impregnate the women and girls of the vanquished villages. They also raped, married or sexually enslaved a phenomenal number of women. Genghis' grandson Kublai Khan, the emperor who welcomed Marco Polo to China, presided by day over the Mongol Empire at its greatest extent. By night he serviced a harem that housed 7000 women, methodically rotated through his bedroom at the time of their greatest fertility.

Immense wealth and power, passed on to male descendants, laid the foundation for the greatest genetic dynasty our species has ever produced. But Genghis Khan's legacy is no freak historic footnote. Men who waged war and sought conquest have long enjoyed high fitness. Men went to war with rape and pillage on top of the agenda, returning with some combination of wives, slaves and treasure. As armies grew in size, the higher a man's military status, the more he profited from a campaign. He then turned these profits into daughters who could be married off to form alliances (and produce grandchildren) and sons who could repeat the pattern of conquest and reproduction.

Violence is a consequence of polygyny, as well as a cause. The fact that a single man can sire a Khan-like number of children means that many other men will have zero evolutionary fitness. It is difficult to overstate the consequences of millennia of sexual selection for avoiding being one of these losers. Remember that our ancestors were the men who strove for wealth and power, and especially those men who strove hardest when the differences between winning and losing were greatest. As a result of this long history of selection, the greater the inequalities between men in a society, the harder they strive. And all this striving leads to competitiveness and violence.

Research on homicide by Martin Daly and the late Margo Wilson is possibly the most thought-provoking body of work to come out of the young science of evolutionary psychology. By analyzing homicide statistics, mostly from the United States and Canada, they showed that by far the highest risk factor for being either a perpetrator or a victim of homicide in these societies is being a young man. The same is true in every society for which data exist. Daly and Wilson's extensive analyses of homicide statistics in modern and historic societies show that more than 95 percent of same-sex killings within societies are men killing men, and that these killings are best understood as "rare, fatal consequences of a ubiquitous competitive struggle among men for status and respect."

Economic and social circumstances affect how desperately young men struggle to gain status in society, recognition from women and respect from rival men. Men are more likely to kill one another in societies and neighborhoods with big wealth inequalities. Early studies of this link were plagued by the problem that inequalities in income are greatest in areas where most people are poor. Perhaps poverty, rather than inequality, is responsible for man-on-man killing. In Canada, though, the relationship between inequality and average income is reversed because the relatively poorer Atlantic provinces have more generous social welfare support than wealthier but more socially conservative western provinces like British Columbia and Alberta. Daly, Wilson and their colleague Shawn Vasdev showed not only that income inequality predicts homicide rates in Canadian provinces but that average income does not. They also showed that year-by-year changes in inequality predict changes in homicide.

In *Bowling for Columbine*, documentary filmmaker Michael Moore famously showed how much less violent Canada is than the United States by crossing the border and finding front doors unlocked and people less concerned about crime. Moore attributed this to America's zeal for the right to carry firearms, but the higher homicide rates in most American states can also be explained by the greater inequality in household income in those states. The most equitable American states have similar homicide rates to the least equitable Canadian provinces.

Inequality within a society drives men—especially young, poor men of relatively low social status—to act aggressively and take big risks in order to improve their prospects: first to avoid being one of the many zero-fitness males and then to be one of the few men who have historically produced the lion's share of descendants. When there are big differences among men in income and in the opportunity to marry or attract extramarital mates, then competition among men intensifies. Not only does inequality among men lead directly to polygyny, but both inequality and polygyny create the intensely competitive conditions in which violence thrives.

The Latter-day Saints

The history of the early Mormon Church illustrates the conflicts at the heart of polygyny. In the late 1820s Joseph Smith received from the Archangel Moroni the Book of Mormon, inscribed in a language he called "Reformed Egyptian" on metal plates. Most of the tough work of writing the book involved Smith staring at a special "seer stone" that he placed in his hat, and dictating his revelations to various scribes, making Joseph Smith something of a middleman in the writing of the Book of Mormon

Soon after the publication of the Book of Mormon, God also revealed to Joseph Smith that some Mormon men would be allowed to marry multiple wives. Smith, his lieutenant Brigham Young and a select handful of powerful Mormon men dutifully took it upon themselves to ensure that this revelation regarding "plural marriages" was upheld. Smith is reputed to have married more than 30 women, and Brigham Young had 51 wives. But the relationship between Mormonism and polygyny has never been a simple one. Smith vehemently denied supporting or engaging in plural marriage himself. Although by some accounts Smith was married or "spiritually sealed" to between 33 and 60 women, his original wife, Emma, denied his polygyny and vehemently opposed the practice until her own death many decades after Smith died.

Smith and other elders went to great lengths to conceal their plural marriages and spiritual sealings to multiple women, some of whom were married to other men. But Smith's secretive polygyny and his authoritarian style provoked anger and resentment within the church, sparking his assassination in 1844 and a subsequent schism that split the church three ways. Brigham Young famously led the largest segment on a westward trek, finally settling what is now Utah. Under Young, plural marriage became an accepted part of church doctrine, but one that only a very few men got to practice. Plural marriage festered within the church and fueled resentment from the rest of America for several decades until it was officially abandoned.

Early church leaders used their positions of influence to further their own reproductive self-interest in much the same unscrupulous ways that leaders of modern-day cults often do. In 1989, David Koresh granted himself exclusive sexual access to all women in the Branch Davidian faction that he led, fathering at least 15 children, some with girls below the age of consent. Jim Jones is said to have banned extramarital sex among members of his People's Temple (1955–78), reserving for himself—and avidly exercising—the right to have sex with both men and women within the cult.

Like David Koresh and Jim Jones, the elders of the early Mormon Church made full use of their positions of spiritual influence to entrench a divinely sanctioned economic and sexual kleptocracy, becoming wealthy from the labors of their flock and stocking their personal harems with the daughters and occasionally the wives of their followers. Splinter groups of Mormons today still marry polygynously, but these communities remain prone to schisms, just like most cults where leaders monopolize the opportunities to mate. Extreme inequality in sexual opportunities is seldom stable for long; it lasted mere decades in the case of the Mormon Church and only years for Jim Jones and David Koresh.

Not all wives are created equal

Polygynous marriage takes hold in a society where men are rare or where the interests of wealthy and influential men win out over the interests of average men, but what about the interests of women? George Bernard Shaw was remarkably prescient when he claimed that "the maternal instinct leads a woman to prefer a tenth share in a first rate man to the exclusive possession of a third rate one." When differences in wealth among men are great, some women might gain access to more resources by sharing a very rich and powerful man with other wives than by monogamously marrying Mr. Average. Some modelers claim that women choosing to maximize the resources they get from their husbands can, alone, explain differences among countries in the proportion of polygynous marriages.

Women might also benefit from sharing a husband if he has great genes to pass on to their children. The anthropologist Bobbi Low found that polygyny is widespread in places where diseases and parasites are rife; especially diseases like malaria, leprosy and leishmaniasis that can kill and that often leave visible marks on survivors. Biologists have found that in various species the best mates are often those who avoid getting infected or who can fight off infections. A mate with good genes for immunity can make a good father for a female's young, by passing those genes on to the mutual offspring.

The idea that women might choose to marry men with great immune genes, even if these men are already married, rather than taking a chance on a man whose immune system is suspect has not attracted the critical scrutiny it deserves. Disease-ridden places tend also to have low average incomes, poor education and dramatic inequality between the richest and the poorest households. Women in these places also have less political, economic and social power relative to men than they do in societies with good hygiene and health services. Every one of these factors should favor the interests of the richest and most powerful men.

To me, inequality and the low status of women in these countries is a more plausible explanation for polygyny than the idea that polygynous women are out to secure the best immune genes for their babies.

But science is not about sifting ideas based on their plausibility. Rather it is about gathering the appropriate evidence and performing, wherever possible, the right experiments to test which idea is correct. I hope that economists and evolutionary biologists will swarm to this issue in the near future, testing whether polygynous men do indeed have immune systems that are genetically superior to monogamous or unmarried men, and whether the benefits of these genes to the offspring exceed the costs that mothers bear from entering into the conflict-infested soap opera of a polygynous marriage.

Perhaps Shaw was wrong and polygyny might not benefit women in general. Sharing husbands generates new and unwelcome conflicts, both between wife and husband and among wives. After all, there are at least two sets of female interests: those of the existing wives and those of the prospective new wife.

A first wife and her husband start out their married life as a monogamous couple, and they share a strong mutual interest in their children. She, he, and their families arrive at the decision to marry in the same way they would in a strictly monogamous society. The trouble begins when he starts thinking about taking a second wife. While this is a good deal for the husband, the first wife has little to gain and plenty to lose. No longer will she and her children have to themselves all the resources and whatever time her husband spends with his family. Instead she will have to share all of this with a newcomer, often one who is younger and more attractive. The decision to take a new wife is clearly an area where the interests of the man and his existing wives can be in enormous conflict.

Who would want to be a second or subsequent wife, destined to share a man's attention and resources with one or more jealous existing wives? Hostility among wives ranges from simmering jealousy to vicious resentment that can boil into violence. Co-wives compete for access to their husband and for material resources, and both conflict and competition can have appalling effects on the health and survival of the children. In sub-Saharan Africa, women in polygynous marriages are more likely to suffer from depression, mental illness and physical abuse from their husbands than monogamously married women in the same societies. And co-wives are more likely to contract sexually transmitted diseases, including AIDS, from their husbands. Most of these afflictions that beset women who share a husband are worse for more junior wives, who suffer lower fertility and whose children are more likely to die in infancy than those of senior wives.

Women, especially women in societies where polygyny is practiced, often don't view polygyny favorably. In the world's biggest polygynous nation, Indonesia, multiple marriages are not popular with women. Indonesian law insists that a man can only take subsequent wives if his existing wives approve and if he can treat them all equally. This law recognizes that men and women have different interests, and keeps polygyny levels modest; only around one in 20 Indonesian marriages is polygynous. Nonetheless, Indonesian women often take to the streets to protest for polygyny to be made illegal, and politicians who are married polygynously usually lose much of the female vote.

But if polygyny is bad for women, then why do many millions of women enter into polygynous marriages of their own accord? Many of these women choose to marry wealthy men who already have one or more wives, and they obtain a good economic deal from doing so. But these women might simply be making the best of a bad situation. Both polygyny and inequitable distribution of wealth occur when a small number of men can secure obscene wealth via conquest, exploitation or inheritance. In these situations, women seldom have the same opportunities to participate in the cash economy and they often cannot inherit or even own land and property. Under these circumstances, some women may find that their best option is to trade their fertility for a part-share in the wealth of a rich and powerful man rather than exclusive marriage to a poor one.

So even when it involves choice, polygyny is often quite coercive. But coercion is not always that subtle. A great many women are forced into polygynous marriages by their fathers, mothers or other relatives, sold for cash or traded for influence and status. Polygyny arises not only from inequities in wealth and power among men; it is also a symptom of great imbalances of power that favour men at the expense of women.

The future of polygyny

Today, more than 50 countries, mostly in Africa and the Muslim world, legally recognize polygyny. Tribal law permits polygynous marriages in much of sub-Saharan Africa. In most of North Africa, the Middle East, and other Asian countries with large Muslim populations, polygynous marriages are sanctioned by civil and Islamic law. Even in Hindu-dominated India, civil law allows men from the sizable Muslim minority to marry up to four times, yet it prohibits polygyny for all other groups. Almost everywhere else, polygyny has been illegal for at least several hundred years. Why have these countries outlawed polygynous marriage?

The first possibility is that legal polygyny is incompatible with mature democracy. More than a century ago, George Bernard Shaw observed that "any marriage system which condemns a majority of the population* to celibacy will be violently wrecked on the pretext that it outrages morality." He had the then-recent Mormon experience in mind when he observed that "polygamy, when tried under modern democratic conditions . . . is wrecked by the revolt of the mass of inferior men who are condemned to celibacy by it." Late in the twentieth century, biologists picked up on this idea. Richard Alexander argued that it becomes

* By which he meant "the majority of men in the population."

ever more difficult for wealthy polygynous leaders to retain the political support of lower-status men when those men cannot marry and reproduce.

Polygyny is, in the long term, incompatible with a smooth-functioning democracy because it promotes the deepest evolutionary interests of the wealthiest and most powerful men at the expense of all other men and all women. Despotic and bloody rule allowed kings and emperors of old to amass phenomenal wealth, marry prolifically and keep harems. But wherever circumstances have made it more difficult for despots to rule, the elites found it easier to gain the loyalty and support of their subjects if those subjects had the stake in society that marriage and family brings, and if the elites were themselves visibly monogamous. Most of the countries where polygyny remains legal are countries where democratic governance, if it is present at all, has only recently superseded dictatorship or monarchy. We can predict that in those countries where democracy matures, the state will cease to sanction polygyny.

The instability of polygynous cults, from the early Mormons to the latter-day Branch Davidians, supports the idea that in the long run, men cannot take many wives and yet retain the political support of other men at the same time. And polygynous politicians' unpopularity with women voters in modern-day Indonesia suggests that women's political franchise is a further nail in the coffin of polygyny under democracy. In the same way that the alleged short-term infidelities of Bill Clinton, Rudy Giuliani and John Edwards dented their appeal to women voters in the United States, polygynists will find it tougher to win elections when women have the vote.

Democratization may tend to kill state-sanctioned polygyny, but even within societies where polygyny is not legal, the richest and most powerful men still achieve a kind of de facto polygyny by keeping mistresses, and by divorcing their wives and marrying younger women. Time a man spends with his mistress, and money spent on her—or on their mutual children—is time and money not available to his wife and legitimate children. But the conflict between wife and mistress goes two ways. Mistresses are worse off than wives in much the same way that the most junior wives suffer in polygynous marriages. Mistresses lack the legal rights enjoyed by wives, including guarantees of continuing material support when the man dies or they break up.

Having a mistress or serially marrying and breeding with fertile women also has the effect of taking women off the mating market, reducing the pool of mates available to other men. This resurrects the evolutionary conflicts of interests between wealthy men and ordinary men. As a result of the polygyny-like conflicts that mistress-keeping and serially monogamizing resurrect, it is not surprising just how vehement many people are in their scorn for the men involved, and often for "kept women" and "home wreckers."

Wealthy and powerful men might concede the evolutionary bonanza of polygyny in order to reinforce their political position, but there are probably other reasons beyond the mere counting of electoral votes for why democratic nations tend only to allow monogamous marriage. Any transition that evens the distribution of power and wealth within a society not only increases the stake that most people feel they have in that society but it also improves the genuine prospects of a great majority of individuals. Success then becomes more dependent on the talents and competitiveness of individuals than on the inheritance of wealth and privilege. It is at exactly this point that parents should invest more in each child, ensuring that they are healthy and sufficiently well educated to succeed in the workplace, attract a good-quality mate, and raise children of their own. The economic changes that accompany democratization make it more important than ever in human history for both parents to invest heavily in their children.

As we saw in chapter 4, parents only have a finite amount of time, money and attention to lavish on their kids and so any increase in the costliness of raising a child means that parents will tend to have fewer children. When children get more expensive, men and women are also forced to rely more on one another's labor, economic contribution to the household and parenting efforts. Polygyny should become less tenable as the economy turns from favoring a select few who gain wealth and power via force and birthright to favoring the men and women who are smartest, hardest working and most ambitious.

This second evolutionary scenario evokes the Catholic Church's own narrative for the existence of marriage. Apparently St. Augustine claimed that the sacrament of marriage was developed to force men to take interest in their children. For a long time, responsibility for abandoned children—especially bastard children—in European and other Christianized nations fell largely to the church. By enforcing marriage, obsessing over bastardry, and pontificating over chastity, the church whittled down the number of children born to single mothers or otherwise abandoned into its so-called care. I am no apologist for religious institutions, especially when it comes to marriage, the family and the welfare of women, but by making monogamous marriage the norm, the church may well have dampened—for a time—some of the evolutionary conflicts of interest between men and women and between men and their children.

Bring me my machine gun

I wrote this chapter not to judge polygyny, polygynists or even Jacob Zuma. I wanted to show how individual men and women are naturally capable of all manner of domestic arrangements, and that the evolved cooperative conflicts between men and women combine with economic circumstances to shape society-level patterns of sex and marriage. I also wanted to show how societies, by the ways they recognize some kinds of marriages as legitimate and desirable, can profoundly alter the lives of their citizens. Polygynous marriage seemed an obvious and sufficiently widespread example with which to do so. Old traditions allowing or even venerating polygyny are not harmless; there are good reasons why it is in the interests of most individuals and the societies in which they live and why polygynous marriage is not sanctioned or supported by most states.

It is no accident that Jacob Zuma's sex and family life draws frequent comparisons to a soap opera. As Zuma's wealth and power have waxed, he has gathered more wives. For some of these wives it may have seemed to be a good deal to marry a man whose political and financial fortunes were on the rise. But instead of allocating his rising income to taking better care of his existing offspring and wives, Zuma has systematically added to his harem. When his third wife, Kate, committed suicide in 2000, she described her 24-year marriage to Zuma as "bitter and most painful," imploring Zuma to take care of her children: "you must not let them starve since I'll be gone, pay their school fees to enable them to further their studies . . . secure guarantee [for] the apartment . . . for my kids to stay without any eviction orders." It appears that much of Kate's agony might have stemmed from Zuma's failure to contribute to their children and meet his obligations as a father.

Zuma defends his polygyny, saying, "There are plenty of politicians who have mistresses and children that they hide so as to pretend they're monogamous. I prefer to be open. I love my wives and I'm proud of my children." This is undoubtedly true, but the question that affects the well-being of more people is not how Zuma feels about his wives and children, but rather how they feel about him and about one another.

Whether Kate alone, among his wives, felt that Zuma had not met his paternal obligations remains unclear, but her bitter unhappiness at Zuma is not surprising in light of what we know about polygynous households. Men who marry many wives, from Indonesian Muslims to American Mormons, often have trouble treating their wives humanely, rather than as slaves or possessions. They also seem to have trouble treating their wives and the children of those wives equally, and one wife often becomes a romantic favorite. The jealous rivalries that anthropologists document among wives of polygynous men appear to be playing themselves out in the Zuma household. Zuma's fourth wife,

Nompumelelo Ntuli-Zuma, is said to be particularly jealous of the attention Jacob has been paying to his fifth and most recent wife, even going so far as to shove her out of the way, in full public view, at the opening of parliament.

So what if Zuma's private life is a soap opera and the lives of his wives and children are diminished by his polygyny? The personal costs are clearly significant, but they are most relevant to the people directly involved. As a defender of traditional polygyny and the tribal sex roles that it reinforces, Zuma legitimizes a practice that hurts many men and nearly all women involved. It brings out the worst in men; the tendency to compete recklessly and violently, the tendency to treat women as reproductive slaves, and the tendency to take less of an interest in each child than their mother does. As a man who repeatedly has affairs outside of his prolific harem of wives, and whose understanding of power dynamics between men and women are as distorted as they appeared in his rape trial, Zuma is making many of South Africa's problems worse. In a country where the incidences of rape and violence against women are among the highest in the world, Jacob Zuma's public lifestyle is a symbolic part of the problem. It also brings into question his ability to make decisions that are in the interests of most citizens.

Even though most people do better when they live in societies that ban polygyny, that should not be seen as an unequivocal endorsement for monotonous monogamy or any other particular arrangement. Romanticizing lifelong monogamy, opposing no-fault divorce and projecting expectations of everlasting monogamous love also harms people who find themselves unhappily married or unable to conform to the expectations of their society, their family or their spouse.

No matter how squeakily monogamous a society is, people in committed relationships—often people deeply in love with their spouses or partners—do seek out extramarital sex. In Victorian Britain, where monogamy and prudishness were elevated to a high art, Arthur Schopenhauer noted that "in London alone there are 80,000 prostitutes.

What are they but the women, who, under the institution of monogamy have come off worse? Theirs is a dreadful fate: they are human sacrifices offered up on the altar of monogamy." Simply because a society decides only to sanction monogamous marriage erases neither men's incentive to mate with many women nor women's reasons to mate with more than one man.

Extramarital activity is no surprise from an evolutionary point of view, and while it can be devastating to the people involved, it need not always be as pathological or destructive as conservatives and religious zealots seem to fear. As Western men and especially women have freed themselves from total economic interdependence over the last century or so, our mating habits have become more fluid. Shorter marriages, if people get married at all, easier divorces for both men and women, and more freedom for young people to enjoy sex might not be as modern as they look. In these respects, we are beginning to resemble the most ancient and egalitarian hunter-gatherers.

8

Where have all the young girls gone?

Every observer or resident of the Indian scene has personal knowledge of at least a handful of homes where childbearing has continued mercilessly until a son has been born, even when this means a string of five, six, or even seven and more daughters first. On the other hand we would be hard put to find even one home which has as many sons and a youngest daughter . . . with whose birth childbearing comes to an end.

— Alaka Basu, 1992

When one sex becomes rarer than the other, selection rapidly corrects this imbalance. But parents sometimes bias their investment in raising sons and daughters. An evolutionary take on sex ratios provides some insights into the atrocity—common in some societies—of neglecting women and girls, killing newborn daughters and aborting female fetuses. It can also help us understand why the excess of young men in these societies threatens regional and global security.

ECONOMIST AMARTYA SEN IS A living demonstration that the dismal science has a conscience. His career-long devotion to understanding the economics of poverty, famine, gender inequality and welfare earned him both the nickname "the Mother Teresa of economics" and the 1998 Nobel Prize. In 1990, Sen drew worldwide attention to the fact that in many societies women are valued so little and suffer such neglect and abuse that they live much shorter lives than men. He estimated that at least 100 million women, mostly in the Indian subcontinent and China, who should be alive had they had access to the same care, food and rights as men, were dead. His analysis, with its focus on the value of women and girls, hardly touched on the question of newborn and as-yet unborn girls, but 20 years later it is clear that since Sen's controversial article at least 100 million more girls and young women are missing because they were either never born or they died very soon after birth.

We know that so many girls are missing because in some countries the ratio of boys to girls is improbably high. In China, for example, in the first five years of this century 120 boys were born for every 100 girls. That may not sound dramatic, but China's population is so vast that by 2020 there will be between 30 and 40 million more boys under 20 than girls in the same age group. And the problem is not confined to China; parts of India, Pakistan, Korea, Taiwan, some former Soviet states and the Balkans also have serious shortages of women and girls. Shockingly, such distorted sex ratios across large populations can only mean that millions of female fetuses and baby girls are being killed.

Aborting female fetuses and killing newborn girls exacerbates a problem that has existed in some societies for centuries, in which girls were sometimes killed but more often neglected or even abandoned, skewing the sex ratio toward boys. These practices are so widespread, and have occurred for so long, that we cannot write them off as mere quirks of foreign cultures or distant eras. In this chapter, I explore how evolutionary biology can help us to understand the origin of strong preferences for sons as well as the deep conflicts of evolutionary interest that can lead to something as repugnant as female infanticide. I ask you to suspend judgment while seeking to understand the world, but not to find unacceptable behavior any more acceptable as a consequence of the understanding gained. I will show not only that our evolved biology can interact with culture and economics to generate misery for millions of people, but also that by understanding biology we can hope to reverse that tide of misery.

The story that I tell involves millions of individual tragedies, from the abortion of female fetuses, and the neglect and death of young girls, to the widespread oppression of women. But apart from these many individual tragedies, the society-scale consequences of biased sex ratios could soon become tragedies on a continental scale as young men with little hope of starting a family of their own reach maturity. These men will be alienated, poor, and very angry. And they will fall into the worst kinds of associations with other young men of equally poor prospects. They will be, in a word, trouble. The problems that the young men of China and south Asia will soon present to those regions and to the world are, again, partly biological in cause. An understanding of evolution is an essential tool in defusing and dealing with those problems.

The problem of having too many men, relative to women, may get much worse if it becomes legal to use new technologies like sperm sorting and embryo screening together with in vitro fertilization (IVF) to routinely choose the preferred sex of children. In Australia and several other developed nations, some owners of IVF clinics and others with a stake in the success of these methods are already lobbying to legalize the routine use of these technologies, allowing parents to choose a child to match the color of the decor in their baby room. As we shall see, evolution predicts that those with the means to pay for IVF-mediated sex selection are precisely the people who are most likely to use these methods to choose boys, but it is the sons of poor people who will pay the price when they mature.

Sex determination

About 105 boys are usually born for every 100 girls. Young boys are slightly more susceptible than girls to disease, accidents and death from violent causes in adolescence. So by early adulthood the numbers of boys and girls evens out. This might seem like a happy accident in which the sex ratio conveniently guarantees one man for every woman, but from another point of view this arrangement is awfully wasteful. If sex were merely about the perpetuation of the species, one man could—and no doubt eagerly would—fertilize the eggs of hundreds of women.

Conservation biologists interested in immediately bolstering the number of a rare species of mammal will often introduce more females than males: more uteruses equals more offspring. Animal breeders know this too. The best stallions cover dozens of mares in a single season at stud, a rooster services hundreds of hens, and semen from the best bulls is used to inseminate thousands of cows. But natural selection is not like a deliberately planned conservation exercise. The winners and losers in evolution are usually individuals, and there is a very strong incentive for most individuals to produce even numbers of sons and daughters, most of the time.

To understand why human sex ratios tend to be roughly even, consider an imaginary population where there are nine women to every man. Because every baby has one biological mother and one biological father, every man would sire nine babies on average for every one baby borne by each woman. This huge fitness bonus for men would mean that genes that bias the chances a baby would be born male rather than female would experience very strong selection, spreading quickly through the population and restoring it within a few generations to an even sex ratio. This potent form of natural selection is called negative frequency-dependence because it always favors genes that bias the chance of conceiving offspring of the rarer sex and away from the commoner sex.

The failure to produce a male heir has ironically condemned more than one royal wife to divorce or execution, and many less-than-royal wives to equally capricious treatment. Ironic because as any high school biology student can now tell you, a baby's sex is due primarily to genes a child inherits from its father and not its mother. Each egg made by a mother contains just half her genetic information. The same is true of each tiny sperm cell a father makes. When a sperm and egg come together in the act of fertilization, the newly conceived embryo has a full complement of genes; half from mom and half from dad.

Whether the baby is a boy or girl depends on whether it gets a normal-looking X chromosome or a miniscule Y chromosome from dad to go with the X chromosome that moms always put in the egg. The Y chromosome carries only a tiny amount of genetic information including the genetic signals that tell the body to make a male. When the successful sperm carries an X, the embryo (with an X from both parents) follows the normal developmental pathway and becomes a girl. When the successful sperm carries a Y to go with the X from mom, this Y alters development resulting in a baby boy.

Because half the male's sperm cells carry Y chromosomes and the other half X, it would seem inevitable that half of all babies would be boys and the other half girls. But the sperm lottery that decides whether a Y- or an X-bearing sperm gets to fertilize the egg can be rigged. For example, the timing of sex relative to ovulation can influence, slightly, whether a Y-bearing or X-bearing sperm wins the race. The acidity of the woman's vagina and her blood glucose levels are two other likely candidates. These effects are complex and subtle, but they have evolved to service our evolutionary interests. Genes that predispose parents to conceive boys are favored when men are scarce, such as during or after a war. Those same genes are selected against when men are too plentiful. The result is a self-adjusting balance that keeps the sex ratio very close to even.

Tipping the scales

Sex ratios tend to be even, but that does not mean individual parents should always produce even numbers of sons and daughters. The development of sophisticated thinking on how animal parents might benefit by investing differently in sons and daughters provides one of the greatest success stories of modern evolutionary biology. In a lecture on sex ratios in 1973, Bob Trivers, at 29 already one of the most influential thinkers in evolutionary biology, mentioned that in many societies a woman often marries into a family wealthier than the family of her birth. Because women tend to marry upward, many men from poorer families and a corresponding number of women from wealthier families are left without anybody to marry. Dan Willard, a PhD student attending the lecture, realized that any tendency for wealthier parents to produce more sons than daughters and for poorer parents to do the opposite could be favored by natural selection. Importantly, such a tendency would be stable as it would leave the overall sex ratio even. To explain Trivers and Willard's idea properly I need to take a short detour into the biology of wasps and red deer.

In species where big females lay more eggs or have more live young than small females, it can pay parents to invest a lot of care in daughters, but not as much in sons. In economic terms, the marginal value of investing in daughters exceeds the marginal value of investing in sons. In many wasps, mothers kill or paralyze an unfortunate insect or spider and then lay a single egg inside its body. When the egg hatches, the body of the prey item is all that the larval wasp has to eat until it becomes an adult. If the prey is small, the emerging adult wasp is small, but large prey items result in a big wasp.

Because of the way that sex is determined in wasps, a mother wasp has absolute control over the sex that each egg will develop into; if she fertilizes the egg it will be a female, but if she leaves it unfertilized the larva will be a male. Big daughters are profitable to the mother because they can lay 20 times as many eggs as small daughters. Big sons do a little better than small sons, but nowhere near as dramatically so, and so a mother tends not to fertilize the egg when she has caught a small prey item, resulting in a small son. When she catches a large prey item, however, she fertilizes the egg and makes a daughter because she gets many more grandchildren from giving a large prey item to a daughter than to a son.

Mammalian males usually grow bigger than females, and the biggest males get the lion's share of matings. In many mammals, mothers in good condition have more sons than expected by chance and mothers in poorer condition have more daughters. Most red deer hinds, for example, can reproduce successfully, but only the biggest and strongest stags achieve the status required to attract a harem of hinds and hold off challenges by other males. It pays red deer mothers who are in top condition to have sons and invest heavily in nourishing them in the womb and by feeding them plenty of milk after birth. And this is exactly what they do; dominant hinds in good condition have sons about 60 percent of the time. This extra investment in sons yields bumper profits if those sons grow big, strong and old enough to hold a harem. Mothers in poorer condition have daughters and do not feed them as much milk because it does not take as much effort to raise a successful hind.

More than a thousand separate scientific studies have tested Trivers and Willard's hypothesis in mammals. Elissa Cameron looked over all these papers and found that when mammalian mothers are well fed and in good condition when they mate, they bear more sons than daughters, but that the opposite is true for mothers in poor condition. She suggested that high levels of circulating glucose might favor the growth and implantation of male embryos. Studies of how cattle embryos conceived by in vitro fertilization (IVF) grow support this idea; male embryos grown in a glucose-rich medium thrive whereas female embryos actually do better when the medium has less glucose.

Human mothers and fathers who are in good physical condition, wealthy, of high status or politically well connected tend to bias care toward sons. Likewise, poor and low-status parents often favor daughters. In modern-day Rwanda, as in many other polygynous societies, second and subsequent wives do not get the support and resources that first wives and women in monogamous marriages do. As a consequence subsequent wives suffer more from stress and their children grow more slowly and are more likely to die in childhood than the children of first or only wives. These lower-ranking wives have more daughters than sole wives or first co-wives, supporting the idea that the mother's status and body condition influence sex ratio at birth.

With an eye for a high-profile data set, Cameron and Frederik Dalerum dug up information on 866 billionaires listed by *Forbes* magazine. They found that billionaires have an average of six sons for every four daughters—as extreme as the most distorted regional sex ratios in twenty-first century China and India. They also looked back at 14 fortunes that had been in the family for more than two generations, and showed that the original billionaire had many more offspring through his sons than through his daughters. Those sons marry fecund wives, and many divorce them and marry new equally fecund wives in a pattern of serial monogamy that might as well be called polygyny. They also presumably have more children through extramarital affairs. Billionaire heiresses still have only one uterus each, and all the money in the world cannot give their fecundity the same boost it gives their brothers. So having sons when you are sitting on a billion or more dollars is a very sound evolutionary strategy, whereas having daughters does not add too much.

Neglect in pre-industrial Germany

I am certain that billionaires are no better nourished than mere millionaires or the comfortably well-off, so whatever mechanism causes the son-bias in billionaire couples might be triggered by extreme *relative* wealth. I imagine that billionaires are just especially likely to conceive sons, and that they aren't doing away with daughters, either before or after birth. In previous centuries, however, wealth and the fatal neglect of daughters often went hand-in-hand.

Parents can discriminate between sons and daughters in small, subtle and often unconscious ways. Slight differences between sons and daughters in how much food parents give them, whether they get immunized, or how quickly parents seek medical care when they fall ill can change the chances of sons and daughters flourishing or dying. Added up over whole populations, even small differences in care can result in dramatic differences in the number of boys and girls dying, distorting the sex ratio.

Detailed records of family life kept by clerics or bureaucrats in pre-industrial Europe have proved a data gold mine to some clever evolutionary biologists. Eckart Voland has spent decades analyzing the records kept in a few German parishes, including Leezen, near Hamburg. Between 1720 and 1869, the chances of sons and daughters living until their first birthday depended on their parents' socioeconomic status. The highest mortality was among sons of the poorest parents—the landless farmers, laborers and tradesmen. The next highest mortality was, surprisingly, among daughters of the wealthy landowning farmers. Sons of landowners were less likely to die before their first birthday and daughters of the landless families were the least likely to die.

In Leezen, daughters commonly married upward into wealthier families. The daughters of landowners had to compete on the marriage market with daughters from less wealthy families, whereas their brothers were spoiled for choice. In the poorer families with no land to bequeath, sons had limited marriage prospects and yet daughters enjoyed a good chance of marrying into a wealthier family. Even though, on average, sons and daughters are equally good at producing grandchildren, wealthy sons are a far better bet than wealthy daughters. And by the same reasoning, poor daughters are a better bet than poor sons. The high mortality of daughters in wealthy families and sons in poor families fits perfectly the logic of Trivers and Willard's idea in which parents adjust their investment in sons and daughters in relation to their long-term evolutionary prospects. Voland thought the same, arguing that daughters of wealthy landowners were probably slightly more likely to be neglected, underfed, and to receive less medical attention than sons, whereas the opposite applied in the poorer landless families.

Interestingly, when landowning parents died, farms most often passed into the ownership of sons, reinforcing the link between wealth and reproductive success within the male lineages of landholding families. From a parent's point of view, passing land on to a son guaranteed many more grandchildren than passing the land to a daughter would have done, and this advantage amplified with every generation in which the land passed down to a son. Voland's studies of another German parish confirm that differences in mortality between infant sons and daughters are tied to how wealth and property change the fitness payoffs of having sons or daughters.

At roughly the same time as the wealthy burghers of Leezen were neglecting their daughters, the opposite was happening about 120 miles away in the parish of Krummhorn. Unlike in Leezen, the Krummhorn population had reached a limit and could not expand. With no new land to farm, there was very little economic growth and few chances for families to build wealth. Wealthy landowners could not leave the farm to several children without impoverishing them. As a result, only one child, usually the youngest son, inherited the farm and raised a family on it. Daughters married, but unlike in Leezen they often married downward into the poorer classes who worked smallholdings or were landless. Other sons, remarkably, often remained unmarried and childless. These sons, it seems, preferred to maintain their social status at the expense of their reproductive success.

The much more competitive economic and ecological conditions in Krummhorn turned the evolutionary payoffs of having sons or daughters upside-down. Landowners had the highest sex ratios at birth (around 116 males per 100 females*), possibly because evolved Trivers–Willard mechanisms biased births toward sons under conditions of relative wealth. But young sons of the wealthy were the frailest children in the whole study and by one year old, sex ratios were even, plummeting to around 91 by 15 years of age. By contrast, sex ratios stayed balanced among the offspring of smallholders and landless laborers. Once again, the wealthy landholders appear to have cared for sons and daughters in dramatically different ways, matching the likely fitness returns from sons and daughters, albeit in the opposite direction from the Leezen parish. The link between property inheritance and the fitness returns to parents of sons or daughters reverberates through historic and contemporary human societies. Interestingly, patterns of inheritance and the practice of women marrying upward are also at play in contemporary China and India.

One-son policy

The most spectacularly male-biased sex ratios in the modern world occur in China where strong preferences for at least one son have meant that for centuries some daughters have been killed in infancy, abandoned or neglected. Even a tiny distortion in the sex ratio of the world's most populous nation means the birth of many millions of extra boys,

* I follow the convention of reporting sex ratios as males per 100 females, so high sex ratios are male biased.

but China's problem is aggravated by history's biggest experiment in social engineering: the one-child policy.

One out of every six people living today is in China, and the population of 1.3 billion grows by 1 million every six weeks. In 1979 the Chinese government mandated that couples could only have one child, implementing Orwellian measures to encourage and coerce citizens to curb their fertility. These measures range from propaganda to reproductive micromanagement of contraception by local leaders, to forced sterilizations, to financial bonuses for adherents and fines for transgressors.

The policy is not implemented uniformly across China, however. It is most extreme in cities, where child-rearing is most expensive and fertility lowest. Among rural families, slightly more laid-back versions of the policy allow some scope for second and even subsequent children. In the densely populated eastern coastal provinces about 40 percent of couples are allowed a second child, usually only if the first is a girl. In the central and southern provinces, every couple can have a second child if the first is a girl, is disabled, or if the parents are themselves both only children. The most remote and sparsely populated provinces such as Inner Mongolia, Tibet and Xinjiang allow all couples to have a second child and sometimes even a third, irrespective of the sex of their existing children.

The fact that one-child regulations sometimes allow families with only girls to reproduce again indicates government acceptance of the importance of sons in Han Chinese society. Children, particularly sons, provide and care for their elderly parents. Bearing sons secures parents a guarantee of care in old age. A daughter is far less certain to provide the same kind of care because when she marries she joins her husband's family.

Allowing urban couples to have only one child, and other couples who already have a daughter to reproduce again, should not alone skew sex ratios. After all, the second child still has roughly the same chance of being a boy or a girl—at least at conception. The one-child policy and the desire for sons can, however, influence what happens to male and female embryos once they implant in the mother's uterus and become fetuses. In the late 1980s and early 1990s, prenatal ultrasound became widely available in China. Not only does ultrasound allow effective diagnosis of many problems with fetuses and placentas, but it makes early sex determination easy and affordable. Combine ultrasound's ability to detect an offspring's sex with the already widespread availability and acceptance of abortion—courtesy of the government's zeal for population control—and you have a recipe for sex-specific abortion.

Although it is illegal in China to abort a fetus purely for reasons of sex determination, the vast majority of fetuses aborted after sex determination by ultrasound are female. Together, these technologies have made it easier for many couples to realize their strong preference for a son. As a result, birth sex ratios have climbed steadily since the 1980s. But abortion is not the only way to eliminate a daughter. Female infanticide is an ancient practice that is both illegal and widespread. In *Message from an Unknown Chinese Mother: Stories of Loss and Love* the Chinese author Xinran captures the heartbreak of dozens of mothers whose daughters have been killed, given away or abandoned. Some mothers live on in quiet misery. Others choose suicide: it may be little accident that China has suicide rates that are among the highest in the world, especially among rural women. Infanticide of newborn daughters is so widespread that it is one of the services that midwives perform for an additional fee. One midwife divulged to Xinran that she routinely strangles an unwanted daughter with the umbilical cord as she emerges, ensuring that the child is recorded as a stillbirth.

Urban Chinese couples tend to have slightly, yet significantly, more boys than girls (115 boys per 100 girls were born in Chinese cities in 2005), suggesting that some couples are using sex-specific abortion in their first pregnancy. If they have one shot at having a son, they will not leave it to chance. The skew is not especially dramatic, however, because urban couples are often less constrained by traditional values, and strong preferences for a son are becoming less socially acceptable.

Son preferences are strongest in the agriculturally productive and often more traditional central provinces where all couples can have a second child if the first is a girl, and some couples even have a third child. Jianxi province, for example, is a relatively poor rice-growing landlocked region in the southeast. In 2005, 137 boys were born there for every 100 girls, which is extremely high, but among first births the sex ratio was only 108 to 100—very slightly higher than normal. The sex bias is driven by the second and third births to families, in which 178 and 206 boys were born for every 100 girls, respectively. Where families know they will have at least a second chance to have a son, few choose the sex of their firstborn. The stakes are much higher at any second or subsequent births, each of which may well be the last opportunity to have one son, and it seems few families leave the sex of their second or third child to chance.

India

When asked about their ideal family size, Indian parents often express a wish for at least two sons. The desired number of daughters varies, but never exceeds the desired number of sons. Indian couples don't just wish for more sons than daughters, they base their family planning strategies on these wishes. Couples that use contraception, and men who undergo voluntary sterilization, only tend to do so after the birth of a second healthy son.

As in China, the strong preference for sons in India has long been a cause of female infanticide, the neglect of young daughters and even the neglect and abuse of adult women. In some parts of India, the chronic neglect and infanticide of daughters was documented as early as the late eighteenth century and may have already been common for centuries before then. In India, as in China, the problem has also been worsening for more than 20 years as ultrasound and abortion have become more readily available.

The preference for sons is greatest in the densely populated north Indian states and territories like Uttar Pradesh, Bihar, and Punjab. According to the 2001 census, most of the 28 states and seven union territories in India have more males than females. Some of the biases are staggering, including those in the territories of Daman and Diu (141), Chandigarh (129) and Delhi (122), and states like Haryana (116) and Punjab (114). And the problem is worst in the youngest cohort of children—in all of these areas the 2001 census counted more than 125 boys for every 100 girls aged one to six.

Son preferences emerge partly—and often in exactly the way Trivers and Willard predicted—from the caste hierarchy that governs the social life and marriage patterns of India's Hindu majority. Brides can marry into a higher, but never a lower, sub-caste. Men from higher sub-castes are a scarce resource because women from both their own and lower sub-castes are effectively competing for them on the marriage market. As a result, the groom's family negotiates a dowry payment, converting the demand for their son into an economic benefit. The higher the groom's caste, the greater his value and the bigger the dowry payment becomes. With a bigger dowry, a family can secure greater upward mobility for a daughter, and thus for any children she has.

Although dowries are now formally illegal in India, they remain a common part of marriage customs. Dominique Lapierre poignantly fictionalized the crippling expense imposed by the dowry system in *City of Joy*, in which Hasari Pal—forced from his subsistence farm by crop failures—struggles as a rickshaw operator to raise sufficient dowry for his eldest daughter to marry well. Hasari dies on the wedding day, destroyed by tuberculosis and a crippling workload, having raised the final part of the dowry by selling the rights to his body for medical research. Instead of working themselves to death, or saving for an extortionate dowry, many would-be parents—especially more affluent ones for whom a dowry payment portends potential financial ruin—opt not to have daughters at all.

As in Han Chinese society, an Indian son, his wife and their children often care for the son's parents. Daughters, who move away when they marry, seldom provide this kind of care. Yet the cost of raising a daughter is even higher in India than it is in China because of the need to invest so heavily in a dowry. The problem is worst in northern India where agriculture and the economy are more traditionally male dominated, marriage is more strictly patrilocal and dowries are bigger than in the south.

In the north, a sub-caste in one village will not accept brides from the same villages to which they send their daughters to be married. Thus a young bride has no kin in her husband's village, and her new neighbors have no links to her natal family. The lack of kinship ties between a young bride and her neighbors reduces her social power, forcing her to rely on the kindness of strangers. Isolated from her genetic relatives and from established allies, a young bride is surrounded by people whose motives and interests are very often at odds with her own. She spends her married life building social alliances and a group of close kin—starting with her own sons and daughters—from scratch. In this context it is understandable that mothers of adult children value the ongoing kinship and allegiances of their sons, who remain in the village when they are old enough to marry, rather than daughters, who leave.

Marriage patterns in south India differ in important ways from those in the north. Villages in the south tend to reciprocally swap marriage partners, creating a much stronger network of genetic relationships and personal allegiances—trade and favors owed—between the bride and groom's family than in the north. A young bride is less isolated from her natal family because she already knows many people in her new village and may be related to several of them. These closer links between married daughters and their natal families have a strong effect both on how girls are valued and treated in childhood, and on the social status enjoyed by women once they are married. In southern India, the pattern of marrying upward to men of higher status sub-castes is also less pronounced, the dowry system is less rigid and extortionate, and women enjoy more powerful and prominent roles within society, farming, and the economy.

It is reasonable to expect that the most affluent Indian city dwellers will have modern attitudes regarding gender, and the lowest incidence of female feticide, as it is known in English (or in Hindi, *khanya bhronn hatya*, "the killing of young girls"). Surprisingly, the opposite is true; feticide is often most extreme in this group. The system of patrilocal marriage is, of course, much more pronounced in rural villages, but even in cities, especially those in the north, sons more often live with or near their parents and support them in their old age. The preference for at least one son is strong, even among the most educated city-dwelling elites. In the countryside, where fertilities remain relatively high, families often keep reproducing until they have one or two sons purely by chance. The urban wealthy, in the thick of the demographic transition, producing fewer babies and investing more in each, have fewer chances to produce a son.

Add to this the effect that dowries are often most extortionate among the upwardly mobile, educated city dwellers. With fewer children, it is more important than ever that each makes the best possible marriage, maximizing upward mobility. The dowry custom—and all its noxious baggage—is spreading from the high castes to social strata where it never existed before. According to Michelle Goldberg, "in recent years, with India suddenly awash in consumerist bounty, dowry demands have exploded, turning the whole thing into a materialistic free-for-all." For many expectant parents the relative benefits and costs of raising sons and daughters simply do not add up, and doctors advertising "Pay 5000 rupees today and save 50,000 rupees tomorrow" are making a killing.

The fact that the children of wealthy city dwellers are the most male-biased group in modern India may be simply due to economic and social factors: the affordability of screening and abortions, enormous dowry expectations, and declining fertility providing fewer opportunities to have a son. But the trend also echoes Trivers and Willard's predictions that when women marry up the socioeconomic hierarchy, the wealthiest groups should have the most male-biased sex ratios. Perhaps, as in preindustrial Germany, urban Indian couples are responding to the marriage prospects of their children. These are precisely the families whose sons will be sought after as grooms and can be expected to attract a large dowry payment. The daughters of these families will have far fewer prospects than the sons, and any match will need to be lubricated with a fat dowry.

Although the wealthier families contribute most to the production of too many men, it is ironically not the sons of these families that will suffer the costs. These relatively wealthy sons will always have good marriage prospects. It is the sons of the poor that will be unable to obtain wives, and that is where the biggest problems for societies begin.

The value of a wife

Just as supply of and demand for goods influences their market value, so the relative availability of men and women who are interested in mating influences the value of each sex in the mating marketplace. When adult sex ratios deviate from even, individuals of the rarer sex become more valuable and the more common sex becomes less so. In economics, when the supply of an item increases or demand for it drops, competition between suppliers brings down the price. On the mating market it is competition between members of the more common sex that raises the intensity of sexual selection within that sex, usually with unhappy consequences.

Throughout history some big deviations in sex ratio have dramatically transformed societies and cultures. Even small deviations from equal sex ratios can cause men and women to adjust their behavior in fascinating ways. In the United States after World War II, the baby boom raised the number of births from 3 million in 1945 to 4.4 million in 1960, after which births fell gradually until the mid-1970s. Because women tend to marry about three years younger than men in the United States, the average woman born in 1948, for example, came onto the mating market about the same time as the average man born in 1945. So for the baby boomers, the women coming onto the market each year were from bigger, younger cohorts than the men.

Female-biased sex ratios in the American baby-boom cohorts marrying between 1965 and 1980 have been blamed for a steady decrease in marriage rates and an increase in divorce rates over that period because an undersupply raised demand for men, giving them less incentive to commit to marriage in the first place, to remain sexually faithful within marriage, to contribute financially, to work in the home and to take a direct hand in raising the children.

Female-biased sex ratios are often associated with increases in teenage pregnancies. Girls become sexually active earlier when they outnumber boys. Having sex younger and more often may be a competitive strategy to keep a boyfriend interested. When the supply of boys dwindles, there are more girls competing for boys' interest and when demand grows, prices are sure to rise. Girls have more sex than they would have had if boys were more abundant, and teen pregnancy is the relatively rare symptom. Likewise, women sometimes find that in an extreme man-drought their best chance of reproducing is to go it alone, further contributing to the rise in unpartnered mothers when men are scarce.

But being in the majority is not all bad news for women. The social scientists Marcia Guttentag and Paul F. Secord suggest that the excess of women, as baby boomers matured, played a role in the rise of second-wave feminism. Compared with the stay-at-home wifely stereotype of the 1950s, far more women in the 1960s and 1970s needed to support themselves than had been the case for the generations immediately before. Baby-boomer women also had to compete sexually with one another for men and to decouple sex from the ever less likely institution of marriage. So the liberalization of sexual mores during the sexual revolution might not only have been to escape the stifling sexual repression of an older generation's values. It seems odd to claim that a demographic quirk produced something as profound and important as feminism's second wave, yet the strange interaction between biology and economics created by the baby-boom–fueled surplus of young women may have propelled a movement whose time had come.

As the ever smaller post-1960 birth cohorts matured, and older baby boomer men divorced and remarried younger women, sex ratios shifted back from the baby-boom–fueled excess of females. From the 1980s until beyond 2000, available men outnumbered women. Under these circumstances men value marriage more highly and are more committed to their families. Today, in US cities where women outnumber men, young men are less likely to get married, preferring to enjoy the opportunities for short-term liaisons created by the glut of women. Men over 35, however, make use of their scarcity by being more likely to marry than when sex ratios are even. Where men are relatively common, they commit to marriage early, competing more intensely to attract a wife in the first place.

In 1910, western states like Montana and Arizona had high ratios of men to women (111 and 110 men to every 100 women respectively) because more men than women had migrated to those states during their settlement. The sex ratio in eastern states was much closer to even. In 1910 wealthier American men were more likely than poorer men to marry, but in the more male-biased states the gap between married and unmarried men in socioeconomic status was much larger. This probably made for a large number of poor men who had almost no marriage prospects, just what we are now seeing emerge in India and China. And the excess of men might be what made the Amercian west such a wild and lawless frontier.

Too many men

Having too many men about the place is not a good thing. For anybody. When the long-established killing and neglect of girls created a large male surplus in nineteenth-century China, men drifted into criminal gangs that terrorized landholders and peasants. These gangs sometimes aggregated into larger militias that visited even greater havoc on those around them. One consequence was the Nien rebellion, which killed more than 100,000 troops and civilians and devastated the economy over a 15-year period.

Just as polygynous societies are more likely to go to war with neighboring groups or tribes, so societies that have too many men tend to throw their weight around. Because young women tend to marry up the socioeconomic hierarchy, young men from the poorest families and those who are uneducated or less employable have the lowest bargaining value in the marriage market. Not only do these men have little stake in society in the form of existing wealth, they also have little prospect of improving their evolutionary fitness by starting their own family. Unable to find wives—and sometimes even work—these men move in groups from one place to another, anonymous transients without the kinds of ties and obligations to their communities that usually prevent antisocial or criminal behavior. These are exactly the kinds of circumstances that make men highly prone to risky, violent, illegal and often unpredictable activities as well as gambling, using alcohol, drugs and prostitution—factors that further increase the risk of antisocial, violent and criminal behavior.

Political scientists Valerie Hudson and Andrea den Boer argue that these problems are on the rise in China, India and wherever else sex ratios are male biased. Kidnapping and trafficking of women are on the rise and the provinces and states with the most male-biased sex ratios tend to have the highest rates of violent crime. The culprits are mostly unmarried young men, often in gangs. Hudson and den Boer calculate that the excess men in China and India will number about 60 million by 2020, threatening not only peace and order within those societies but also regional and global stability.

The problem of enormous numbers of excess men cannot easily be undone. It must, instead, be managed. As the wildly sex-biased cohorts of children in places like India and China reach adulthood, the excess young men with poor reproductive prospects will make a lot of trouble. I predict that theft, assault and homicide will rise catastrophically, matched only by abduction and trafficking in women for marriage, prostitution and sexual slavery. New approaches to preventing and managing crime and violence are going to be needed urgently.

So are new approaches to international relations. The extent to which surplus men in countries like China, Taiwan, India and Pakistan tip the belligerence of those societies remains to be seen, but this is potentially a problem that should concern us all. The concern is even greater when many of the young men with few reproductive prospects are within the grasp of Pakistani or Afghan madrassas or other places where poverty and alienation mix so explosively with fundamentalism.

Reasons for hope

Countries like India and China are far from culturally uniform, and the distortions in sex ratio vary among regions and even among neighborhoods. Often marrying outside their own ethnic group provides a happy resolution for those individuals worst-off in the mating markets. An article in *The Economist* magazine highlights the case of Baljeet Singh, a Hindu truck driver, and his young Muslim wife, Sona Katum. Singh, from the north Indian state of Haryana (sex ratio 116), is one of an increasing number of men from states with highly male-biased sex ratios who are bucking the caste-based orthodoxies and making previously unthinkable matches, including marrying Muslims. Katum —from a group that does not have the same male-biased sex ratios—has enhanced her own prospects, escaping extreme poverty in her home district by making what for her family only a generation ago would have been an equally scandalous match.

The practices that give rise to distorted sex ratios can often be self-defeating. When one sex is heavily in demand, that demand is often met by immigrants from other cultural groups, like Sona Katum. This kind of mobility might eventually erase some cultural differences between groups, including many of the practices that generated the bias in the first place. The sex-ratio–driven changes in the economics of mating and marriage also tend to ensure that dowries or bride prices do not continue to escalate forever. Where women are in short supply, they become more valuable. In places like southern Africa, where polygyny creates a shortage of women, men are often compelled to pay a bride price. Although in India recent economic development and explosive consumerism have seen both a spread and an escalation of the dowry system, the sheer demand for brides should eventually arrest the cultural momentum behind the dowry problem.

In Haryana and other parts of India where women are in acutely short supply, this is already happening. In addition to marrying women from previously unthinkable groups, the onus appears to be shifting toward men to pay a bride price and often to foot the expense of finding and relocating a wife. In China many ethnic minorities do not share the strong son-preference of the Han Chinese, and are not as strictly subject to the one-child policy. We can expect that women from these groups will increasingly marry into Han families, with the potential to weaken some of the cultural preferences for sons.

Natural selection can restore sex ratios by favoring any mechanisms that bias sex ratio toward daughters. Over the next two decades there will be so many more men than women in India, China and elsewhere that women will experience much higher average fitness than men. If some of those women have genes that bias the sex ratio toward daughters, we may find a measurable shift toward daughters. But this kind of change takes many generations. If current practices continue for centuries, we can expect that in areas that experience strong male biases, a very high proportion of embryos and fetuses will be female. Let us hope that we will have found easier ways of overcoming the social and economic damage of having too many men by then.

As Amartya Sen pointed out in his essay, the state of Kerala does not have the same sex-ratio problem as the rest of India. In the 2001 census there were 95 males to every 100 Keralese females, and there have been more women than men in Kerala for at least a century. Kerala is different from most of India in a few important respects. Women's life expectancy exceeds that of men by six years, due at least in part to accessible healthcare facilities that are unrivaled in any other state. The literacy rate of 99 percent is by far the highest for any Indian state (the national average is 65 percent), and this reflects a history of state-funded education that traces back before the early nineteenth century. Kerala also has the best welfare system, a history of left-wing and often communist governments, and a citizenry that is more politically engaged than those in the majority of more conservative states.

Sen also draws attention to the fact that the Nair caste to which about one-fifth of Keralese belong tends to trace its lineages through the mother's line. Property is inherited more equally among brothers and sisters. The Nairs have long been very influential in this area, having ruled and administered most of the small kingdoms that now make up Kerala. The importance of women in both family history and the inheritance of wealth is likely to influence not only the status of women within the Nair caste, but also more broadly in Kerala. The modestly female-biased sex ratio in Kerala suggests not only that girls and women do not suffer more than sons from the chronic neglect and lack of access to healthcare that typify much of the rest of India, but also that the practices of sex-biased abortion and infanticide are uncommon in Kerala. Clearly societies that value, educate and empower adult women also value daughters.

Male-biased sex ratios are a symptom of a society that does not value women enough. Anything that biases sex ratios toward men means that the value of women on the mating market increases, but whether this can be broadened into a more general increase in how societies value women's lives and contributions depends on what societies are prepared to do about it. South Korea had, a generation ago, a similar preference for sons as well as male-biased sex ratios to rival India and China. That country curbed both its population growth and its preference for sons by massively improving the education of girls and the participation of women in the economy. While sex ratios have been climbing over the last two decades in India and China, they have fallen to far less distorted levels in South Korea in less than a generation. The South Korean success story remains incompletely understood, but it shows that the great momentum generated by evolutionary and cultural forces and their powerful interactions can be overcome.

9

Blame it on the Stones

After silence, that which comes nearest to expressing the inexpressible is music.

—Aldous Huxley, 1931.

Music is the quintessence of culture but it also has evolutionary roots that run deep. Music stimulates parts of our brains that first evolved for other purposes, but music-making has also evolved in its own right. It is especially important in courtship, and in learning to navigate the tricky social transition to adulthood. It may be the most complex and sophisticated courtship display in the animal kingdom.

THE ROLLING STONES ARE THE biggest, baddest rock-and-roll band of them all.* Those of us born in or after the 1970s find it difficult to grasp just how huge the Stones were in the 1960s, and how chaotically insane the world around them must have been. Hysterical swarms of screaming fans plagued them at every public appearance, a constabulary intent on

* They may not have sold as many records as The Beatles or AC/DC, but they outlasted The Beatles by decades, and they were international megastars when the Young brothers were still wearing school uniforms.

stamping out drugs by stamping on the Stones hounded them relentlessly, and just about every public commentator derided them scornfully. At the time, they were so aggressive, lascivious and degenerate they made the mop-topped smiley-faced Beatles look definitively clean-cut. But to an evolutionary biologist, the Stones exemplify a problem that screams out for an explanation. Why bother with all that rock-and-roll nonsense? Brian Jones, the talented multi-instrumentalist and flamboyant early driving force in the band, took way too many drugs, and became increasingly difficult to work with and unable to contribute to recordings and live performances. He aggressively fell out with band members and was eventually fired. He was soon dead; drowned—possibly murdered—in his swimming pool at the age of 27. Being a rock star did not accord Brian Jones a long and productive life.

The remaining chapters of this book revolve around popular music, especially rock and other Western popular music since the 1950s.* It is tempting to dismiss music as a purely cultural phenomenon and rock as an aberration of twentieth-century culture: an arbitrary set of attitudes and chord changes stolen from the blues and passed from one rocker to another, modified only by inspiration. That is pretty much the kind of explanation that musicologists and cultural theorists have long favored as they dissect the arcane details of who-copied-what-from-whom, interpreting the history of popular music in a mumbo-jumbo of postmodern critique.

By contrast, an evolutionary biologist starts from the assumption that things don't happen by themselves, especially things that dramatically increase your risk of dying, as rocking out certainly does. Anything as popular, exciting, sexy, deadly and—most of all—as difficult to do well, needs an explanation. But we need to ask the right questions. We know rock is only two generations old. Asking how making and listening

* I'm not here to split hairs about musical history: I use rock 'n' roll, rock and roll, and rock interchangeably to mean the broader musical movement.

to music affects the reproductive fitness of individual musicians and audience members can only tell us part of the story. We also need to consider other evolutionary processes that have operated on individuals and their genes that might predispose something as sexy and dangerous as rock to shake, rattle and revolutionize the modern world.

I want to debunk not only the idea that rock and other musical genres are exclusively cultural phenomena, but also the idea that there are any phenomena that are so exclusively cultural that our evolved biology is irrelevant. No sane person would argue that rock is not cultural: it is well known that rock arose in the 1950s out of existing musical traditions including rhythm and blues, folk, blues, jazz and country. It spread through learning and imitation, assisted by a special blend of social and economic circumstances that arose soon after World War II and the spread of technologies like commercial radio, record players and television. But even though rock is a quintessentially cultural phenomenon, it grew in the soil of our evolved biology. That is what makes it so utterly compelling, and why, a decade into the twenty-first century, it is still going so strong. Lesser art forms like disco, tantra, Morris dancing and macramé may tap into aspects of our biology, but they don't have the same alchemic blend of sex, rebellion, anger, danger and freedom. That is why they are lesser art forms. Nothing short of that opium of the masses, religion, with which rock shares so many ritual similarities, even approaches rock music for cultural expression of raw human biology.

Cake

Before there was rock there was music, and the origin of music is an area where evolutionary biologists have already had plenty to say. Harvard University neuroscientist Steven Pinker, a great champion of the adaptive view of human nature, argued in *How the Mind Works* that music is not really an adaptation but rather that it is like cheesecake. Strawberry cheesecake, for example, is a modern invention that tickles our evolved tendency to enjoy

> the sweet taste of ripe fruit, the creamy mouth feel of fats and oils from nuts and meat, and the coolness of fresh water. Cheesecake packs a sensual wallop unlike anything in the natural world because it is a brew of megadoses of agreeable stimuli which we concocted for the express purpose of pressing our pleasure buttons.

Pinker goes on: "music is auditory cheesecake, an exquisite confection crafted to tickle the sensitive spots of . . . our mental faculties."

Pinker makes an important point. We should always be cautious when claiming that something is an adaptation that has been shaped by natural selection. We can understand how our evolutionary past incidentally made the brand new invention we know as cheesecake irresistible without claiming that the ability to make cheesecake is an adaptation. Natural selection builds on the raw materials that already exist, and our minds probably evolved some of their capacity to enjoy rhythm and predictability, and to be soothed or aroused by certain sounds and voices long before anybody made anything we would now call music. But that does not mean that the music we enjoy today is simply a clever trick that pushes our pre-existing buttons in the same way cheesecake titillates our pre-existing tastes and food preferences.

For a start, music, unlike cheesecake, is no newcomer to human society. All contemporary peoples make and enjoy music, and this suggests that musicality arose well before modern humans spread out of Africa. The oldest known musical instruments are bone flutes more than 30,000 years old, but people probably sang, clapped, and banged sticks and stones musically long before they ever made and played instruments. Music is also an important focus in the traditional lives of contemporary hunter-gatherers. If people have made music for tens and possibly hundreds of thousands of years, then selection had plenty of chance to shape our music-making. We should seriously consider this possibility before swallowing whole the idea of auditory cheesecake.

Until the invention of the phonograph in the 1870s, people played and listened to music live. It was almost always a highly social art form that involved groups of people. Contemporary tribes make music and dance as inclusive groups. Churchgoers trudge onward together through their weekly hymns, like Christian soldiers. Troops march to stirringly patriotic tunes from military bands. In almost any city, on any night of the week, women and men dance in clubs, pubs and live music venues. And the biggest crowds of all are those at rock concerts. In 1969, three days after Brian Jones' death, the surviving Stones played a long-planned free concert to 250,000 fans in London's Hyde Park. This number wilts alongside the unfathomable 3.5 million people who braved heat, humidity and fear of crowds* to attend a Rod Stewart concert at Rio's Copacabana beach on New Year's Eve 1994.

Perhaps, then, music's role at the heart of social-group living might be important to understanding its evolution. Might it be a good way for a tribe to bond, dissipating petty conflicts and improving the functioning of the group? Or might it make the group stronger, stirring bravery and a strong sense of belonging before and even during battle with other groups? These ideas seem intuitive; what else could explain the invention of the bagpipe?

But the idea of group benefits tends to make evolutionary biologists squirm. For a trait to evolve via the benefits it delivers to a group, it must cause some groups to thrive or others to wither away. The problem here is that while it is good to be on the winning side of a conflict, it is even better to be a member of the winning side who was slow into battle and didn't get hurt or killed. Selection in which some individuals do better than others is usually more potent than selection in which some groups prevail over others, so we should look very carefully for individual-level benefits before we buy wholeheartedly into the idea of group-level benefits.

* Crowds composed of people, I might add, who like Rod Stewart concerts.

While selection might indeed favor music because it works wonders on groups of people, there are many plausible benefits of being an accomplished musician and of enjoying listening and dancing to music. Music may have arisen from a song-like communication system that evolved for courting mates, much like the independently evolved song systems of whales, gibbons and birds. Ever more sophisticated and meaningful sounds, and the ability to decipher them, gave the most articulate hominids in each generation an advantage in snagging and seducing the best mates, leading eventually to language. This elegant idea was first suggested in 1871 by Charles Darwin in his second great book on evolution, *The Descent of Man, and Selection in Relation to Sex*.

According to Darwin, "it appears probable that the progenitors of man, either the males or females or both sexes, before acquiring the power of expressing their mutual love in articulate language, endeavoured to charm each other with musical notes and rhythm." Darwin arrived at this idea because he had been thinking about features of animals and plants that seem unnecessarily costly, like the songs and bright plumage of many birds, which make the bird conspicuous to predators. Darwin recognized that these traits improve an individual's mating opportunities and he predicted that this mating advantage causes bright and beautiful traits to evolve via sexual selection.

Sexual selection unleashes some of the most rapid and extreme evolutionary change in the animal world because reproduction is the currency of evolutionary success. Attractive signals and the preferences for those signals influence one another's evolution, pushing both signal and preference to outlandish extremes. The animal world teems with courtship signals; crickets chirp by night, moths puff cocktails of irresistible chemicals into the night air, manakins flash their brightly colored feathers in the dappled forest light, and fishes lurking in turbid rivers pulse electric fields to find and attract mates.

For Darwin, music was the midwife at the birth of language, and our capacity to speak evolved from our capacity to make musical sounds.

What we now know as music and what we now know as language have since evolved from this single origin. It may never be possible to disentangle the chicken-and-egg problem of whether language or music came first, but it seems that the two depend on some of the same hardware. Many of the parts of the brain that we use to create and understand language are also involved when we make or listen to music. But there are also differences in how music and language stimulate our brains, suggesting that music and language, though related, are not one and the same thing.

Despite having been around for 140 years, Darwin's ideas on the evolution of music and conversation via sexual selection have only received the attention that they deserve from a small number of authors. The evolutionary psychologist Geoffrey Miller is one of those few. In his entertaining and erudite book *The Mating Mind*, Miller argues that the human mind is at least as much an elaborate organ of courtship as it is a tool of survival. According to Miller, some of the human mind's most impressive features have evolved to attract and entertain potential mates, persuade them to have sex, to settle down and to raise a family.

These include our human gifts for language and conversation. From establishing first contact to the point at which they decide to have sex, a couple are almost always in conversation when they are together. Conversing in an interesting and engaging way with the opposite sex is so difficult that it takes most of us years of painful trial and error to learn—if we ever manage it at all. Even when we are confident and competent enough to ask somebody out, the surest sign that a date is going wrong is when the conversation dries up. Likewise in long-married couples, a loss of interest in conversation by one or both parties can be just as devastating to the relationship as a loss of interest in sex. As Germaine Greer put it, "Loneliness is never more cruel than when it is felt in close propinquity with someone who has ceased to communicate."

Sing more, get more

If conversation is difficult, Miller argues, the rhyme and meter of poetry are infinitely more so—and thus poetic ability is more impressive than mere eloquence. Singing well, which requires not only timing and rhythm but a command of tone, melody and harmony, is an even more impressive achievement. The ability to write songs that combine all of these elements and that lyrically ring true is a precious miracle of nature. Beyond the rather difficult business of playing, singing and writing music, the music and lyrics themselves may also be forms of courtship.

It turns out that Brian Jones left a clue to the evolutionary problem posed by his early death. He fathered four children, each by a different mother. The other Stones who did not share Jones' misfortune did mostly share his talent for proliferation. Stones' guitarist Keith Richards, who to everybody's amazement is still alive and playing at 68 years old, had five children with two different women. Vocalist Mick Jagger, twice married and famously philandering, has been linked in five different decades with some of the most desirable women on the planet, officially siring seven children by four different women.

Although drummer Charlie Watts breaks the hyper-sexualized mold, Brian, Keith and Mick personify the fossilized footprint of a long history of sexually selected music-making in our species. None of these men can count themselves among the most prolific fathers of their time, but who knows what heights of paternity they might have scaled had not the 1960s also ushered in history's biggest advances in contraception? The sheer number of fertile women who got very close to the Stones, sometimes by the most ingenious of routes, is experienced by very few men in history. Or as Keith Richards puts it, "You stood as much chance in a fucking river full of piranhas."

Musicians overcome two of the biggest evolutionary problems that face people looking for a mate: meeting or being noticed by potential mates, and courting or seducing them. From the day a band plays its first live gig, band members are in the business of exposing themselves to potential mates—whether that is their intention or not. Musicians in small-time bands enjoy a slight advantage over their non-musical brethren in meeting and getting to know possible mates, and the more successful the band becomes, the bigger the audience at their gigs and the larger the pool of possible mates. If ancient musicians enjoyed even a fraction of the mating success of modern rock stars then sexual selection on music-making abilities would have been sensationally strong throughout our evolutionary past.

Men and women from a great many societies use music to privately court one another, from young singers at Saturday-night love markets in the hill villages of northern Vietnam to the evening serenades of eighteenth-century Italy. Music-making, singing and dancing can also be very public displays of prowess. According to Radiohead, "Anyone can play guitar," but a world of pawned guitars, amps and drum kits tells a different story. Learning to play a musical instrument is no trivial task, and only a tiny fraction of those who begin reach any level of proficiency. The fact that so many great rock musicians are self-taught only shows that rock music and the instruments of choice are well suited to let people with limited means and opportunity realize what talent and motivation they have. When all the elements work together we have one of the living world's greatest sexual displays.

The notion of music as a sexual display explains much more than why we bother making music—it also explains why music can be so sublime that its power transcends mere description. Natural selection usually favors sober functionality: teeth that cut and grind food, day in and day out, lasting as long as our ancestors could have hoped to survive; bones strong enough to support an active body but not so sturdy that they become too heavy; and a gut that wrings every last morsel of nourishment from every meal. But attractiveness is different—it keeps on evolving. Genes that make an individual attractive can, in a very few generations, come to be so common that whatever was attractive ten generations ago might be merely ordinary today. As Geoffrey Miller puts it:

> ancestral hominid-Hendrixes could never say, "OK, our music's good enough, we can stop now," because they were competing with all the hominid-Eric-Claptons, hominid-Jerry-Garcias, and hominid-John-Lennons. The aesthetic and emotional power of music is exactly what we would expect from sexual selection's arms race to impress minds like ours.

Close to you

For all the sex, music would not have as much power as it does if it were merely a tool for the talented and inspired few to get their rocks off. For most of our species' past, making and dancing to music involved most, if not all, of the adults and teenagers in the group, much as it does in contemporary tribes and villages. Social occasions gave young men and women many opportunities to observe and enjoy one another's music-making and dancing and to use this information to judge who might be the best mate. Repeated chances to observe members of the opposite sex doing something difficult are exactly the kind of assessment that is most reliable when animals assess the genetic quality of their potential mates.

Music becomes an even more potent part of courtship when it is coupled with dancing. For my grandparents' generation, being able to waltz and foxtrot were essential social graces. Dancing was one of the few reliable ways for men and women to get close enough to converse. Wallflowers who did not dance missed most of the opportunities on offer. Even when I was in college in the 1990s the most popular club on campus was the ballroom dancing society, which ran four back-to-back classes five nights a week. Mastering the *paso doble* or the *tango* remains an impressive and deeply romantic feat, but even informal and improvised dancing can signal the dancer's tenderness, deftness and coordination. Or lack thereof.

Recent studies have used high-speed video to capture the movements of men as they dance, and then to animate computer-generated bodies with those movements. Women asked to watch those animations prefer the dancers who move more vigorously, with more bending and twisting motions of their neck and torso and of their right knee (most of the men were right-footed). Maybe so many men are secretly terrified of dancing because they know that women are using it to assess them. Perhaps dancing is to many men what the bikini is to many women.

As twentieth-century technology made music ever more portable, so it made it possible for mortals to choose the sound track we use for dancing, courtship and seduction. Early in the rock era it became possible for young women and men to signal their taste, politics and personalities by the music they listened to on the radio, played from car stereos and bought for their beloveds. Later, the mixed tape or CD allowed one to put both love and thought into a gift without spending much money or learning to play an instrument. Even though few people today are proficient and persistent enough to master an instrument, most courtships still begin within earshot of a live band, a DJ or a stereo. It is less dramatic and certainly less effective to play a song on your stereo than it is on your guitar, but it is also so much easier. We contract our chosen artists to play the music we use for our own courtship, and we pay them handsomely.

The technologies that gave people the power to choose their personal sound track were the launch platforms for the major developments in twentieth-century music, including jazz, rock 'n' roll and hip-hop. Phonographs (record players) made it possible for people all around the world to listen to the same songs recorded by the same bands. As jazz exploded in the first decades of the century, musicians could become megastars in distant cities they had never before visited.

Radio and television each, in turn, allowed people to choose and listen to their own music in ways that had never before been possible. As post-war prosperity spread through the United States and the middle class ballooned, record players and radios became common household items, and televisions were soon to follow. The average teenager suddenly had more disposable income and came under less pressure than previous generations to grow up, have babies or enter the workforce. They suddenly became a powerful consumer group, buying 45 rpm record singles, feeding jukeboxes and devouring radio programmed just for them.

With hip-hop, again, technology pushed things along. Boomboxes, mixtapes and samplers enabled a new generation to make and remake their own music. Each of these twentieth-century technological advances made it increasingly possible to sell different music to different members of the household. Each generation of teenagers and young adults grabbed hold of the new opportunities, eager to escape the stultifying and straitlaced music of their parents.

Teen spirit

The first album I ever bought with my own money was U2's *Under a Blood Red Sky*. I was 13, and all my high school friends were playing it. Thirteen is a typical age to start buying your own music and asserting your own musical tastes. In early adolescence people stop becoming passive consumers of the music their parents and older siblings are playing, and start forming their own musical tastes. Our sexual and social identities and our musical tastes are shaped in adolescence and early adulthood, and it is no accident that sex and music cozily knit themselves together at this time.

As an interesting thought experiment, quickly think of one of your favorite songs and try to remember when you first heard it. When I did this exercise I immediately thought of "Losing my Religion" by R.E.M., long one of my favorite bands. When I heard it first I was 21, on a holiday with university friends. I was hooked by its unusual mandolin riff and consumed by its dark unrequited longing.* Research on memory shows that people recall more personal events from early adulthood than from later adulthood or childhood—something psychologists call the reminiscence bump. A recent study by Steve Janssen at the University of Amsterdam shows that the bands and songs people remember most clearly and fondly in middle age are the ones they came to love between the ages of 16 and 21—late adolescence and early adulthood.

By contrast, when asked to name favorite books and movies, people favor more recent works. The strong musical reminiscence bump between 16 and 21 years of age makes sense because our relationship with music really gets going when we enter puberty, and becomes most intense from then through early adulthood. This is no coincidence. It is the music that plays when we fall in love, when our hearts break, when we discover sex and learn the meaning of true friendship. It is the sound track to which we find our way through the most important and perilous transition since the day we left our mother's womb and entered the loud, bright, air-breathing outside world.

If you want to know the most important difference between chimpanzees and modern humans, look no further than the speed at which we develop. A newborn chimpanzee has a brain about half the size of an adult chimp's whereas a human baby is born with a brain that is only one-quarter of its final adult size. This is due to the kind of flaw in our anatomy that only a deeply misogynist designer or a blind incremental process like natural selection could arrive at: the fact that

* That was very definitely me in the corner, and in the spotlight.

the mammalian birth canal passes through the pelvis. So, relative to chimps, human babies are born only half-baked, and both brain and child need a longer time to grow and mature. A chimpanzee takes about ten years to become a fully functional adult, but humans only mature in their late teens. A protracted childhood and lingering adolescence are among the most important evolutionary changes since we diverged from the other apes. One likely reason for such a long adolescence is that as our brains—and the complex social systems and sex lives that these big brains enabled—became ever more elaborate, so we needed a longer time to prepare for our reproductive years.

Young boys do not normally grow beards, or muscular upper bodies, and they don't show the frightening aggression and relentless interest in sex that young men do. In fact to contemplate such a thing is shockingly distasteful. Likewise young girls do not normally grow breasts, ovulate or menstruate, and when they do the stories make tabloid headlines. But natural selection tunes the timing of all of these transitions in much the same way it adjusts more obvious traits like the way our stomach enzymes work. Breasts and beards do not start to grow until adolescence because young children have no adaptive use for them; instead boys and girls need simply to grow as fast as they can and to learn as much as they can about the things that are common to everybody, like walking, talking and what to eat. Natural selection adjusts the timing of adolescence so that our bodies usually only grow the features that make us into men and women as we near the age where our adolescent ancestors needed them. Just like beards, breasts and menstruation, our brains develop from those of boys and girls into the brains of men and women who are ready and able to make it socially in an adult world.

The hormone-addled development of a body, and especially a brain, from a standard-issue children's version into the specialized body and brain of an adult man or woman can be terrifying. Not only must adolescents come to terms with the physical changes to their bodies, but they must learn to work with a more varied and more intense palette of emotions. They learn to navigate the complex relationships, obligations and conventions of adulthood. They also learn to make, break and re-fashion alliances with peers of the same sex. And they learn by trial and all-too-embarrassing error the opaque rules of love: how to love and be loved; how to recognize when love is requited and how to stay sane when it is not. On top of this, teenagers have to deal with parents who have their own evolutionary and social interests, and who themselves are making the clumsy transition from coddling a precious child to guiding and letting go of an emerging adult. No wonder those experiencing adolescence seldom have the confidence or the perspective that would allow them to enjoy their precious youth.

It is also as teenagers that we begin our tussle with questions of identity. Who am I? What is my purpose? What do other people think of me? And probably most important: Am I the only one who feels this way? Fashioning a coherent identity is one of the most important issues that we grapple with in our lifetime, and no time in our lives is more important in resolving our identity than late adolescence and early adulthood. The adolescent brain has finally matured enough to keep track of what we think about large numbers of other people and what we think they think about us. It can also begin to grasp—however imperfectly—the differences in how the minds of others work. On top of that, we can begin to understand the question of social status—both the status we inherit due to the status of our parents, and the status that comes from our own talent, industry, and hanging out with the right crowd.

"Music hath charms to soothe the savage breast," and by savage I mean teenage. The right tunes can greatly soothe the savage turbulence of the adolescent years. Adolescents have probably long found music helpful as they discover how to use their full palette of adult emotions, and how to negotiate their increasingly complex social worlds. Despite the often heroic efforts of teachers, teenagers have much more to learn than the things they are taught in high school. Important as Shakespeare, the Krebs cycle and differential calculus are, teenagers are often more focused on another, more ancient, kind of learning: about themselves, about how other people and society work and most of all about love, sex, and the complicated package that comes with it. From Chuck Berry's "School Days" to Bruce Springsteen's "No Surrender," rock and roll has long existed at this crossroads between education and learning, school and real life.

Education research confirms three facts that have been obvious to shrewd writers and musicians for some time: we learn from stories, we learn best when we are having fun, and young people learn much more readily from their same-age or slightly older peers. By these measures, popular music is a fantastic way to learn. Barry A. Farber's entertaining book *Rock 'n' Roll Wisdom* gives a psychologist's view of the many lessons within great rock lyrics, about topics from friendship to depression and from identity to dying. Rock is awash with trite, banal and even nonsensical lyrics, but some of it is wisdom so distilled that it should be sold in small bottles with a big warning label. So much so that great lyricists like Bob Dylan, Patti Smith and Leonard Cohen are also considered great poets.*

In the rock era, teenagers certainly found in music much wisdom, as well as comfort and companionship. Artists are often only slightly older than their audiences, giving the artists both the credibility of youth and the authority of a slight age advantage. When a favorite artist sings about infatuation, rage or powerlessness, it takes little effort to identify immediately with the musician and her message. Who has not felt that a particular song was written only for them? Of course music was not invented mid-twentieth century, and the operas of Mozart, Puccini and Verdi alone probably convey as much about love, friendship and conflict as the entire last decade of popular music. Was Don Giovanni not the archetype for the rock-and-roll bad-boy? But popular music's power is

* I pointedly don't count Jim Morrison here. Anybody responsible for the rhyming of "road" with "toad" in "Riders on the Storm" automatically disqualifies himself as a poet.

its immediacy. Even if the themes are ancient, and the latest incarnation appears inarticulate to some oldies, the important thing is that the audience own and relate to the music. For now who cares what some long-dead Europeans had to say in warbling Italian?

Getting to know you

Put two teenagers or young adults together in a context where they need to become acquainted, and the conversation will soon turn to music. Jason Rentfrow and Sam Gosling did exactly this with 60 undergraduates at the University of Texas, pairing them up on an online bulletin-board system and asking them to get to know one another. In the first week of the study almost 60 percent of participants spoke about music; more than books, clothing, movies, television and sports combined. Rentfrow and Gosling also analyzed the information people post on their Internet dating-site profiles, and it follows the same pattern; the bands and music that people prefer is by far the most commonly reported information. Not only do people advertise information about their taste in music as a way of advertising their identity, but they place great faith in it as a way of forming judgments about others.

In a series of studies, Rentfrow, Gosling and their colleagues show that musical tastes powerfully predict people's personalities (see box on p. 218: The Big Five Personality Factors and the Central Six). By analyzing the preferences of almost two thousand people, they found four major dimensions along which musical tastes vary:

1 *Reflective and Complex* tastes for music, such as the blues, jazz, classical and folk music, tend to indicate that a person is Emotionally Stable, Open to new experiences, and has above-average intelligence and verbal ability.
2 *Intense and Rebellious* tastes, including a liking for rock, alternative

or heavy metal, tend to be shared by people who are Open, athletic, and of above-average intelligence and verbal ability.

3 People with *Upbeat and Conventional* taste tend to like country, sound tracks, religious music and pop, and are usually Agreeable, Extroverted, Conscientious, politically conservative, wealthy, athletic, with low Openness, dominance and verbal ability. (I always knew there was something suspect about pop-lovers.)

4 Last, folks who like rap or hip-hop, soul, funk, and electronica are said to have *Energetic and Rhythmic* tastes, and they are highly Extroverted, Agreeable and athletic, tend to speak their minds, and are often politically liberal.

These associations between broad musical tastes and personalities are anything but arbitrary. People with different personality types seek different types of experience and stimulation, which they tend to find in different types of music. Part of the reason there are so many kinds of music is likely to be found in the fact that there are so many different combinations of personality traits.

Preferring certain bands and even certain songs over others advertises even more precise information. In the 1990s there could not have been a bigger gulf in outlook and politics between politically progressive bands like Nirvana or Rage Against the Machine and the more derivative, bombastic and often antisocial music of Guns 'n' Roses or Bon Jovi, even though to outsiders or less committed rock fans* they would all have sounded like hard guitar-driven rock. The identities and images that particular bands construct are all a large part of their appeal. Like tends to assort with like when it comes to personality types, and fans of a band tend to have personalities that match the personality the band constructs. We advertise our personalities by the music we play and talk about and by the band paraphernalia we wear or display.

* And twenty-first century classic rock radio stations.

THE BIG FIVE PERSONALITY FACTORS AND THE CENTRAL SIX

Personalities are very complex and no two people are identical. Yet it is possible to boil down a lot of the variation in personalities into a few, manageable dimensions. This has long been the goal of personality psychology. The most popular current taxonomy is the Big Five or five-factor model, according to which there are five main dimensions along which people's personalities differ, often named Openness, Conscientiousness, Extroversion, Agreeableness and Neuroticism.

Certain combinations of traits tend to be found together. People who are imaginitive, adventurous and show great insight often also have broad interests, and are curious. People who share all of these attributes score high on the dimension called **Openness**. The other four dimensions are:

- **Conscientiousness** (the tendency to be thoughtful, self-disciplined, goal-directed, organized, and to care about details)
- **Extroversion** (talkative, sociable and highly expressive)
- **Agreeableness** (kind, affectionate, altruistic, trusting, cooperative)
- **Neuroticism** (anxious, moody, irritable, sad and emotionally unstable). Sometimes Neuroticism is called Emotional Stability, but that kind of wrecks the OCEAN mnemonic.

There are many great resources on the Web for exploring the Big Five personality factors, including tests. One of the best is Sam Gosling's site at <www.outofservice.com/bigfive>.

Some psychologists add General Intelligence, the tendency to do well on a variety of tests of cognitive ability, to the list of general personality factors. With the addition of General Intelligence, the Big Five becomes the Central Six.

Musical tastes are good signals because they are hard to fake, should we ever wish to. My own tastes run between *Intense and Rebellious* and *Reflective and Complex*, with a limited smattering of *Energetic and Rhythmic*. I could not fake my way through a conversation about modern hip-hop because any half-devoted fan would immediately sense that I can't tell Trick Daddy from Diddy, or Jay-Z from Jay Smooth. In order to learn enough to hold even a brief conversation in a convincing way, I would have to listen to their music and follow their careers. It is hard enough keeping abreast of my research group's taste for zef-rap.

People advertise those bands with which they identify by playing their records—often at improbably loud volumes in cars—talking about them, reading about them in music and gossip magazines, and wearing their merchandise. The Ramones stopped playing in 1996, and the core members are dead, but Ramones T-shirts are still selling well. Apparently nothing says "I Don't Give a Fuck" more convincingly in the early twenty-first century than a freshly minted Ramones T-shirt.* Some of the best-known logos in the commercial world belong to bands such as AC/DC, the Rolling Stones, KISS, Nine Inch Nails, Guns 'n' Roses and Metallica. The urge to buy T-shirts advertising our favorite bands can be overwhelming, but some fans are so committed they have a band logo tattooed on their skin. These outward signals of music preferences are possibly the single most effective and widespread way for teenagers in particular to identify like-minded peers and potential allies, and are about as reliable a conversation starter as you can get.

As well as signaling personality traits, musical tastes also signal and even confer status. Anti-school and anti-establishment subcultures, stacked with young men, often revolve either around sports or around very narrow single-genre music tastes, like heavy metal, rap and in some places country music. In the early days of rock, low-status rock

* The true descendants of punk are not the replacement Ramones or the surviving Sex Pistols; try looking elsewhere, like Cape Town's Die Antwoord.

subcultures like this were also common. Social alienation can be remedied by retreating into a difficult-to-penetrate world where the chief currency is a detailed and arcane knowledge of a certain kind of music. In obvious contrast, good students with happy social and family lives—the kind of teenagers likely to achieve high status as adults—are more likely to adopt a range of tastes including popular genres like pop, reflective genres like folk or blues and less youth-oriented music like classical and opera. Perhaps intelligence and the social skills that confer status come packaged together with this ability to be a musical omnivore. Or maybe the ability to be an omnivore and to fit in with many specialized groups actually confers status, in much the same way that a politician must often rely on connecting with a range of very different constituencies in order to be elected.

Remembering sex

The vast role that popular music plays in the lives of teenagers and young adults as they begin to navigate the world of men, women, infatuation, love and sex reinforces the fact that music derives much of its power from courtship and mating. But what of the elderly and, ahem, the middle-aged? Some of our strongest memories in middle and old age involve a song that was playing when we first met, danced with, or made love to a special somebody. The reminiscence bump in which we most fondly remember the music of our early adulthood, as we stood on the threshold of our sexual prime, again reinforces the tight link between our musical and sexual identities.

Even well into old age, people report that music helps them to understand and develop their identity, connect with others, maintain their well-being and express their spirituality. This is, apart from some differences in emphasis, pretty much the role that music plays in the lives of the young. The difference is that the middle-aged and the elderly listen with a sense of reminiscence and sometimes of regret for a lost youth, a sense that does not yet encumber the young. That is why we pay big money to attend reunion concerts or see long-forgotten acts from the dark side of our youth. The Australian humorist and one of my favorite writers, Mark Dapin, pointed me to this quote which he believes may apply to all reunion concerts. It comes from a piece by the modern historian Timothy Garton Ash, writing about the audience at a reunion of the sixties Czech pop group the Golden Kids:

> Sometimes they clap along. But when the Golden Kids sing "Suzanne" there's just total silence.
>
> . . .
>
> Tense and heavy with regret: the silence of the middle-aged remembering sex.

10

About a boy

It all started out with the best intentions, and out of total poverty . . . We just wanted to dig in our heels and rock the fuck out.

—Soundgarden and Pearl Jam drummer Matt Cameron, 2010, on the start of grunge

Rock was the sound track to the sexual revolution, yet feminism's second wave seems to have passed it by; rock musicians are overwhelmingly male. This may be a symptom of a sexist industry but it also shows the signature of the processes that generate biological sex differences. In this chapter I will argue that rock music was heavily influenced by the evolutionary agenda of male performers, seeking to out-compete other men and attract women, often by working in tight fraternal coalitions.

IN MARCH 2002, THE BRITISH Broadcasting Corporation launched BBC 6 Music, a radio station for serious lovers of pop and rock music, with the motto "closer to the music that matters." It was a phenomenal success, focusing on popular music since the 1960s, with music experts as presenters, plenty of live sets, unique and archived recording sessions, documentary material and interviews. It also involved listeners in innovative ways, from in-depth online message boards to SMS-driven programming. The listeners were engaged, passionate and very male: so overwhelmingly male, in fact, that in late 2007 the BBC made some changes—including introducing apparently female-friendly "personality DJs"—to boost their female listenership. These changes were about as popular with 6 Music's existing listeners as suggesting *Britain's Got Talent* singer Susan Boyle play drums at a Led Zeppelin reunion.

Attempting to explain the changes, Lesley Douglas, then-head of popular music at the BBC, argued that men and women listen to music in different ways: "For women, there tends to be more emotional reaction to music. Men tend to be more interested in the intellectual side of the music, the tracks, where albums have been made, that sort of thing." Ms. Douglas' comments, as you might anticipate, did not go down well with a lot of people, including many women music journalists.

The BBC 6 Music affair and Lesley Douglas' comments were far from an isolated flare-up. In 2004, *Rolling Stone* magazine celebrated the 50th anniversary of rock 'n' roll by asking a select panel of musicians, writers and industry insiders to vote for the artists that most influenced rock's first half-century. From the tallied votes, they named the Immortals—the 100 greatest acts of all time. Among the 322 solo artists and regular members of those acts, only 26 were women. There were 296 men. The highest-ranked woman, Aretha Franklin, came in ninth behind The Beatles, Bob Dylan, Elvis, the Stones, Chuck Berry, Jimi Hendrix, James Brown and Little Richard.

I would hesitate to claim that *Rolling Stone* manufactured their list out of hostility to female recording artists, although only two of the 52 panelists who voted for the list were women. In fact, the list might underestimate the male bias in rock, with 14 of the 26 women coming from three bands: the Shirelles (76th), Martha and the Vandellas (96th) and Diana Ross and the Supremes (97th). Deserving as these bands are of Immortality, and as important as they were in the last 50 years of popular music, one wouldn't immediately consider them rock bands.

It is very difficult to say anything at all about sex differences, much less use words like "emotional" and "intellectual" as Lesley Douglas did, without being accused of imperiling all the gains of first- and second-wave feminism and unleashing the fury of the people who made those gains. And arguing that evolved biological differences between men and women are partly responsible for inequities like the ratio of rock gods to goddesses risks—according to some critics—legitimizing centuries of sexist oppression. As a result, people who realize that the fight for sexual equality is both important and far from over, and I include myself and a great many evolutionary biologists in that category, fear to tread near the question of human sex differences.

But here's the problem: the sexes are undeniably different in many ways, from the risk of being born autistic to the chance of being executed for murder, and these differences begin with biological differences that can't be wished away. My main reason for taking this journey into the very male world of rock 'n' roll music is to explore the unsafe and unsteady ground of sex differences, how they come about and what they mean.

One argument I will not make is that men and women differ, on average, in their ability to make great music. Girls and boys display similar musical ability in childhood; where they are free to choose, similar numbers of girls and boys take up an instrument and there are no discernible differences in musical talent at primary school. Although

Women involved in *Billboard* number ones (percentage of artists)

DECADE	ALL NO. 1s	ROCK	POP	R&B/ SOUL	DISCO/ DANCE	HIP-HOP/ RAP
1960s	15	1	20	9		
1970s	18	5	20	12	35	
1980s	18	7	27	13	41	0
1990s	32	11	54	10	48	10
2000s	34	12	45	11	48	19

popular music has long been dominated by men, women's involvement rose steadily since the 1950s as societal constraints crumbled. A look at all the number one records since *Billboard* began its charts in 1958 confirms this (see table on p. 224). In the 1960s, 15 percent of the performers were women but this number reached 34 percent in the decade ending in 2009. One in three musicians at the very peak of popular music today is a woman, and perhaps the numbers of men and women will be equal by the 2030s. Women clearly make music that is every bit as good and as popular as the music men make.

But the gains for women came in the genres of disco/dance and pop, not in rock or R&B/soul. Critics and marketing people sort music into ever more arcane taxonomies to impose order on a cacophony of styles and influences, yet for many musicians categories stultify creativity and constrain their opportunities to work. I cannot tell whether the music women make is less likely to be shoved into the rock pigeonhole because of the qualities of the music or simply because it is made by women. But there can be no denying that when rock 'n' roll exploded, the fortunes of women musicians imploded.

Between 1949 and 1953, immediately before rock 'n' roll took hold, women sang 34 percent of all hit songs, but this dropped to 12 percent between 1957 and 1960 as rock, in all its machismo, came to dominate popular music. In the 1960s women made up only 1 percent of musicians on *Billboard* number one songs that are identified as rock recordings, and this number stuttered to only 12 percent in the past decade (see table). Part of the spectacular gain in the number of women on number one hits in the last two decades comes through the demise of rock; where 40 to 50 percent of the music reaching number one in the 1960s, 1970s and 1980s was some form of rock, this dropped to 10 percent by the 1990s and even less in the noughties.

The story that I tell in this chapter, of music periodically being hijacked by angry young men to suit their evolved agenda, is currently playing out again in rap music, one of the chief beneficiaries of rock's mainstream demise. It is a story possibly as old as music itself. It happened at the birth of jazz, the blues, reggae, and might even be what propelled music's first great rock 'n' roller, Wolfgang Amadeus Mozart. It also applies, in parts, to sport, to literature and many other great fields of endeavor. But rock makes a fascinating case study.

La différence

In the previous chapter I argued that the capacity to make music evolved, at least in part, by sexual selection. Even for non-musicians, listening and dancing to music is a near-indispensable part of courtship and learning about love. But men and women tend to have different agendas when it comes to sex, love and courtship; sometimes the differences are small and subtle but sometimes women and men can seem to be as far apart as Venus and Mars. At least they do in the personal growth aisle of the bookstore.

If you want to watch an encounter between people from two very distant planets, take a journey through the literature on sex differences: Mercury, let's say, exports social contructionists, whereas biological determinists are from Uranus. To biologists who study sexual selection on animals it is no surprise that differences in the ways males and females reproduce lead to differences in behavior and anatomy. Yet on the planet Uranus lives a special kind of slow-moving biological determinist who uses big, blunt stereotypes about animals to draw conclusions about what is "natural" in humans. They occasionally come out of their caves to argue that women can't read maps or that men can't communicate using polysyllables. Oversimplifications like these usually belie a lack of imagination, and they also have a nasty habit of being wrong. I come to bury such clumsy explanations for sex differences and not to praise them.

Unfortunately, the evil that is done in the name of an idea does not necessarily make the idea untrue. At the other end of the solar system things are so changeable on Mercury that the inhabitants believe all differences between men and women are due to differences in experience, socially constructed by the stereotyped ways people treat boys and girls. Instead of building a subtle understanding of how we are shaped by the combined effects of biology and experience, social constructionists—often the people we expect to think most deeply about sex differences and how they affect society—have thrown the biological baby out with the sexist bathwater, preferring the plainly incredible position that gender has nothing to do with biology.

Just as radio listeners love an old familiar song, so news outlets and their readers love an old familiar story with a simple two-way conflict at its heart. One song played on high rotation pits unchanging nature against mercurial nurture, evolved genes against socially constructed environment and ancient evolution against contemporary culture. Hard-and-fast biological and wishy-washy social constructionist explanations for sex differences miss, in all their rabid antipathy, the all-important common ground. The tiny genetic switch that makes some embryos male and others female is a *developmental* switch. It sets the embryo, and later the child, adolescent and even adult off on a course of growth, development and learning that is a conversation between our genes, the environment, our bodies and our cultures. These developmental conversations tailor a body or a behavioral repertoire specifically to the conditions in which an individual develops.

Male field crickets call for hours each night to attract females, and in so doing neatly illustrate this idea. My colleague Michael Kasumovic has shown that if you play developing juvenile male field crickets the calls of dozens of adult males, they take their time developing into an adult so that each grows big enough and has the energy stores to compete with all the other males he hears around him. But if you play males the calls of only a few competitors, they become adults two to three days sooner, keen to start calling and attracting females while there are few competitors about. Female crickets do the opposite—they rush to mature when fooled into believing there are lots of attractive males around and they take their time when they think males are thin on the ground. It makes no sense to say that these strategies are genetic or that they are environmental—natural selection has shaped the genes involved in cricket development to respond to the environment in ways that depend both on the number of calling males and the sex of the cricket. In this way, both males and females can make the best out of the situation, but in different ways.

Behavioral differences between women and men also respond to local conditions. In this book I have discussed how economic circumstances interact with evolved biology to shape domestic sex roles, marriage patterns, divorce, birth rates, obesity and even the sex ratio of children. In almost every case, the overlap between men and women is greatest and the average difference smallest when women and men are not constrained in how they can participate in workplaces, the home and society, and where men and women have equal power to negotiate the evolved cooperative conflicts over when to have babies, how many to have, and how much to care for each of them. Economists recognize that the drivers of individual behavior add up to create the large-scale patterns they call "aggregate outcomes." Evolutionary biology, like microeconomics, explains large-scale patterns as the aggregate outcomes of individuals making the best of the circumstances into which they are born. Sex differences are aggregate outcomes of evolved responses to ecological, economic and cultural circumstances. They are neither immutable nor inevitable, but across large numbers of people they are certainly present and persistent.

Yet even when sex differences in a trait persist, such differences provide no defensible justification for sexism. As Steven Pinker puts it:

> There is, in fact, no incompatibility between the principles of feminism and the possibility that men and women are not psychologically identical. . . . Equality is not the empirical

> claim that all groups of humans are interchangeable; it is the moral principle that individuals should not be judged or constrained by the average properties of their group.

The extreme view that differences between men and women are 100 percent socially constructed and that biology is irrelevant goes hand-in-hand with the view that gaining and exercising power is the main motive of human social life. Social constructionists—often under the intoxicating influences of Marxism and postmodernism—see social interactions as groups exercising power over other groups. In this view, all sexism comes from men as a group exerting power over women. The contrast with an evolutionary view could not be clearer. Mostly, natural selection is driven by interactions among individuals and among genes. This is my main point; that cultural phenomena like rock 'n' roll, and broad patterns like sex differences, can emerge from the behavior of large numbers of individuals—all acting in their own approximate interests—even without a coordinated agenda of top-down oppression.

That is not to deny the existence of top-down oppression. In chapter 7, I explored how men who attain great power can act, in their individual interests, to impose a marriage system that oppresses all women and most other men. Top-down sexism has certainly played its part in music too. Consider the record company executives whose decisions determine which bands thrive and which bands wither. This largely male clique has long influenced female artists and interpreted female tastes in ways that pervert women's contributions to rock. "I'm terrified by how (today's top women artists) are all controlled by a male corporate idea of what women and rebels should be. When Christina Aguilera is taken seriously as a rebellious figure, we have a huge problem," wrote Garbage's Shirley Manson, naming Patti Smith as *Rolling Stone* Immortal number 47. Punk pioneer and rock goddess Smith cut her own defiant path through rock and beyond the hard place of industry sexism, but she is one of the desperately few.

But just how much of the gender bias in rock we should pin on record industry executives remains an open question. It is tempting to build conspiracy theories, but fiercely competitive markets operate, like natural selection, from the bottom up. Record industry executives quickly learn, like all astute business people, that a dollar has the same value irrespective of the sex or age of the consumer. The last 30 years of popular music have seen these power brokers finding ways to use female voices, those of Shirley Manson and Christina Aguilera alike, to appeal to new fans, more often female fans, but largely outside of rock.

Maybe the oppressor resides among those who write the history of rock. I don't mean the writers who ghost the stars' own sanitized accounts of their fabulous careers, but the critics and historians who make sense of the chaotic events, influences and people that shaped music. According to musician and music author Elijah Wald, throughout twentieth-century popular music, including jazz and rock, "the critics were consistently male—and, more specifically . . . they tended to be the sort of men who collected and discussed music rather than dancing to it." This, according to Wald, "is relevant when one is trying to understand why they loved the music they loved and hated the music they hated." So Elvis is now considered so much more important than the almost equally popular Pat Boone, and PhD students write theses about The Velvet Underground but not about KC and the Sunshine Band. If Wald is right, then Nick Cave will almost certainly be seen as more important than the infinitely more popular Kylie Minogue; and Jay-Z will totally eclipse his wife, Beyoncé.

These biases among moguls and historians almost certainly reinforced the maleness of rock, pushing male performers and immortalizing the acts that spoke to the largely male clique who intellectualize about it. But that doesn't seem like a satisfactory explanation. Why, for one thing, did largely male moguls and historians feel the way they did about those acts?

The main point of this chapter is that rock is so male biased because men—large numbers of individual men—seized on rock and exploited it to serve their own agenda, giving rock music a masculine shape and voice. That male agenda has its roots in the evolutionary challenges that men, especially young men, have always faced: winning respect and status among their peers, challenging the authority of older men, and attracting women both for long-term relationships and one-night love affairs. That is neither to say these forces are irrelevant to other musical genres nor that women have played no part in rock. Rather, I'm simply saying that men got on the rock 'n' roll tour bus faster and in bigger numbers, recognizing it as the perfect vehicle in which to fulfill their appetites.

It crawled from the South

The music from which rock 'n' roll was born throbbed with more libidinous content and performance than anything coming out of white America at the time. Even the name was salacious. "Rock 'n' roll," according to the cultural historian Michael Ventura,

> was a term from the juke joints of the South, long in use by the forties, when a music started being heard that had no name, wasn't jazz and wasn't simply blues and wasn't Cajun, but had all those elements and could not be ignored. In those juke joints "rock 'n' roll" hadn't meant the name of a music, it meant "to fuck."

When the music crossed over to white middle-class radio, so many songs involved rocking and rolling that radio deejays—possibly naively, perhaps cunningly—called it "rock 'n' roll."

From its very birth, rock sizzled with sex. Elvis was such a sex bomb that his FBI file apparently declared him a "definite danger to the security of the United States," and it only got more dangerous from there. Great rockers have exuded every kind of sexy; the sneering raunchiness of Jagger; the blue-collar machismo of Springsteen; the brooding darkness of Nick Cave; the wit and intelligence of John Lennon; and the tantric boasts of Sting. Prince, for example, lacks the strong, masculine features and tall stature of other regular inhabitants of "World's Sexiest Man" lists, but his musical prowess, smooth moves and outrageous wardrobe have made him a literal sex symbol.

Rock gods don't just happen to be sexy. They are deities because of the sexy way in which they rock, because rock is *all* about the sex. The always-entertaining cultural theorist Camille Paglia gets it right, albeit a little pretentiously: "If you live in rock and roll, as I do, you see the reality of sex, of male lust and women being aroused by male lust. It attracts women. It doesn't repel them." If rock 'n' roll is all about the sex, we are comfortably in the intellectual domain of evolutionary biology. After all, in evolution, as in so many relationships, sex changes everything.

Sex machine

It is difficult right now to recall just how special Tiger Woods used to be. Seemingly overnight he turned golf from a plodding snooze-fest into something exciting and uncluttered by the bumbling preppiness that always encumbered it. By his very presence at a golf tournament, Woods could double television audiences and on-course attendance. But Tiger's original destiny was to transform more than golf. According to his late father, Earl, Tiger was the Chosen One, destined to "do more than any other man in history to change the course of humanity." But since Tiger's prolific extramarital shenaniganizing erupted publicly late in 2009, it looks like—as Monty Python might have put it—"He's not the messiah, he's just a very naughty boy."

Tiger and his public-relations mind-sanitizers patronizingly chose, as his defense, for Tiger to seek rehabilitation for sex addiction. You might infer, from the media coverage and conspicuous displays of indignation by sponsors, administrators and politicians, that Tiger Woods was indeed suffering from a rare and contagious affliction, and that he was the first sporting idol caught playing away games. He definitely was not. The indiscretions that littered the playing career of Australia's greatest living cricketer, Shane Warne, probably cost him the test captaincy, a role considered more prestigious and influential in Australia than prime minister. Even *The Greatest*, Muhammad Ali, proud Muslim, conscientious objector and an inspiration in the American struggle for racial equality, was a legendary pants man. In the lead-up to Ali–Frazier III, the Thrilla in Manila, Ali flaunted his 18-year-old mistress so brazenly that Imelda Marcos mistakenly called her Mrs. Ali during a visit to Malacanang Palace. The real Mrs. Khalilah Ali, at home with four children, was so incensed at what she saw on television that she flew to Manila to confront The Greatest. Ali's press conference to explain the situation was remarked by one journalist to be "the first time that a major celebrity called a press conference to announce his marital infidelity." It definitely was not the last.

Am I the only person who finds all the posturing and condemnation of Woods—and every straying sports idol before or since—extremely bizarre? Certainly, the private and public humiliation to which unfaithful sportsmen subject their wives, and the way their behavior devastates their young families, are immense private tragedies. Nothing I say in this chapter exonerates them. But it is also hardly surprising to me that men of their fame and wealth, with access to an almost limitless number of starstruck attractive women, occasionally take advantage of the opportunities that fame presents. In fact, for many sportsmen and the adolescent boys who dream of emulating them, access to and recognition from women is one of the great purposes and perquisites of stardom.

Sexual selection acts differently on men and women because of differences in what it takes to reproduce successfully. Sexual selection on men can be especially strong because some men—like Genghis Khan and his sons and grandsons—sire children with thousands of women but many men don't get to mate at all. Modern birth control blurs the relationship between the drive for promiscuous sex and the evolutionary payoffs. Unlike Muhammad Ali, who has six children by two of his wives and two others from extramarital relationships, our unfaithful megastars have not converted their extraordinary appeal into prodigious offspring. At least not as far as the public record goes. Nonetheless, effective contraception is a largely new invention. The urge to mate with large numbers of attractive and fertile women, given half a chance, evolved because men who felt that urge and managed it wisely reaped the ultimate evolutionary reward: prolific reproduction.

Not every man can be Genghis Khan, Tiger Woods or Shane Warne, but that doesn't stop them trying. Men who marry more than one wife —simultaneously or sequentially—or who sire children through their extramarital activities, end up in the "winners" column of the evolutionary score sheet. Those men are our ancestors, and the celibate, impotent or sexually uninterested men who lived alongside them are not. Their unsexy genes died along with them. Of course evolutionary success takes more than being a raging sex machine, and there are many other songs in men's reproductive repertoires. Being a good father and partner can have its evolutionary benefits, but it's like a steady job with a known income. Hitting the evolutionary jackpot is a high-risk high-reward strategy.

Warrior-kings of old and the modern warrior-sportsmen of today have no stranglehold on sexual conquest. Many of today's successful businessmen, actors, authors and politicians dally discreetly and sometimes —Mr. Sheen—not so discreetly. But since the middle of last century one group has outperformed them all: rock stars. Evolutionary psychologists Martin Daly and Margo Wilson described the harems kept by Genghis Khan's male descendants as "manifestations of male appetites, released from the usual constraints of personal power. A well-guarded harem of nubile women is the realisation of a male fantasy." They might just as well have been writing about rock stardom, which has all the ingredients for sexual conquest on a scale not seen since Genghis Khan: trivially easy access to legions of adoring fans, booze and drugs that lower inhibitions, money that can buy gifts, hotel suites and sometimes silence, and shows that sizzle with sexual energy. No wonder Dire Straits reckoned rock stars get their "money for nothing" and their "chicks" at no extra cost.

Long before a band ever makes serious money, they attract an audience and a following, and the first reward of fledgling success is in the oldest and most basic currency of all. As Gary Herman writes in *Rock 'n' Roll Babylon*:

> Somewhere at the heart of rock 'n' roll's magic is the groupie —whether nameless fan or glamorous socialite—who converts music into the currency of the orgasm and spends freely. She (it is almost invariably a woman) is the central character in the myths and legends of rock—proof to fans and observers that rock stars have found the modern holy grail of guiltless promiscuity.

To Keith Richards, forming the Rolling Stones in 1962 immediately transformed his sex life: "Six months ago I couldn't get laid; I'd have had to pay for it. One minute no chick in the world. No fucking way . . . and the next they're sniffing around."

Although rock elevated guiltless promiscuity to a high art, it is also wonderfully effective at winning over long-term partners. When asked about how he lured Anita Pallenberg—who bore Richards' first three children—away from Brian Jones, Richards professes an ineptitude that would comfort most mortal men:

> I have never put the make on a girl in my life . . . I just don't know how to do it. My instincts are always to leave it to the woman . . . I'm tongue-tied. I suppose every woman I've been with, they've had to put the make on me.

Easy when you're a rock gazillionaire.

The infamous rock 'n' roll lifestyle is, in large part, a modern-day projection of millennia of sexual selection on men. Talented male musicians jumped at the opportunities presented by the economic and cultural circumstances of 1950s America and 1960s Britain, using their musical ability to realize their evolved fantasies. But the link between sex and rock is about more than men's frantic attempts to woo choosy females. Rock also carries the signature of fierce competition among men for status and respect.

Get rich or die trying

Seduction can take more than a song, a poem or a fancy turn of phrase. Throughout history, as far as we can tell, women tended to mate and raise families more often with men of status and wealth. Men gained much more than women did by taking big risks and striving for prominent success. It should be no surprise, then, that men are generally far more likely to engage in risky behaviors and to strive relentlessly for status than women.

But it doesn't pay to strive, compete and take risks under all circumstances. In modern industrialized countries, where men's evolutionary success is not all that variable and women can accumulate their own wealth more than at any other time in history, the sex difference in risk-taking and striving is relatively small compared with the polygynous agrarian societies I discussed in chapter 7. The most important determinant of a man's competitiveness and risk-taking behavior is the level of inequity that surrounds him. The bigger the gap in wealth between haves and have-nots, the greater a man's chances of being one of life's many anonymous male losers who leave no descendants, and the bigger the gap in behavior between men and women.

Dire Straits were being ironic when they sang of "money for nothing." When we envy millionaire rock stars we are staring at the tiny tip of a substantial iceberg of amateur and professional musicians, many of whom never reach the surface. People in the music business talk of artists that have "paid their dues," meaning not only that their peers acknowledge their musicianship, but also that they have been through an extended period of struggle and often poverty rather than, say, being propelled into the limelight by reality television. Far more artists quit during this struggle than ever make it to fame, recognition and wealth on the other side.

This is nothing new either. Louis-Henri Murger wrote in the preface of *Scènes de la Vie de Bohème*, his novel on which Puccini's opera *La Bohème* and the rock opera *Rent* are based, that "Bohemia is a stage in artistic life; it is the preface to the Academy, the hospital, or the morgue." We are accustomed to tales of musicians struggling like the poet Rodolfo from *La Bohème* or Murger himself while they perfect their trade and establish a following in a hostile and corrupt marketplace. The stories we know end in triumph and recognition because these are back stories of the musicians who did, eventually, make it. It is difficult to know how many more quit their musical dream or wound up dead. Or as AC/DC sang early on the long road to eventual success, "It's a Long Way to the Top (If You Wanna Rock 'n' Roll)."

With the spread of recording and playback technologies, the amount of live music that people listened and danced to dropped steadily throughout the twentieth century. Before the phonograph and the radio, most people experienced music by playing together or going out to listen to orchestras and bands. Many more professional musicians made a decent living playing in a local ensemble six nights a week than can make a living today. But few musicians played consistently to such big crowds that they could become ridiculously wealthy. Some of the best money was made by composers and publishers from sales of sheet music—amateurs would buy music to play the popular songs of the day at home, and professional musicians would be expected to play a variety of old favorites and new popular works at their concerts. The technologies that ushered in the rock-and-roll era made it possible for everybody to listen not only to the same songs, but the same versions of those songs played by the same musicians. The musician became every bit as much the message as the music, and this made it possible to become a global megastar, often seemingly overnight.

The result cleaved a gulf between the musicians who hit the big time, and those who did not. The chance to make a respectable living as a competent musician with a relatively normal working life dwindled. The large sector of professional musicians eroded to a smaller sector of struggling musicians hungry for their shot at the big time, and a tiny number of superstars. Reaching the big time is usually a long, hard slog, with no guarantee of ever making it. Those lucky few who do arrive hurtle into a life of relentless touring and playing to large clubs and arenas, bringing wealth and fame beyond the wildest dreams of most.

Just as social and economic inequity brings out aggression, risk-taking, status-seeking and coalition-formation behavior, particularly in young men, I predict from evolutionary theory that increasing inequality in commercial success among musicians is exactly the kind of economic factor that should bring out the risk-taking, highly competitive streak in musicians, especially young male musicians. The rising inequality throughout the history of rock coincided with the emergence of sub-genres of rock dripping with masculine antisociality including ever greater machismo, aggression, and often misogyny.

Hard rock and heavy metal explode with these themes, but there is an independent strand of music with the same tendencies: gangsta rap. The antisocial extremes inhabited by heavy metal and gangsta rap both overdose on the male motivation for status and success, often at the expense of other men. The hardest rock, metal and rap express the timeless alienation and helplessness of young men with poor prospects. Boys and men with good prospects, born into equitable societies, or those born on the right side of the tracks in less equitable societies, don't have to take as many chances to make it. But young men with less favorable prospects live risk-filled lives, with a small shot at making the big time but a much bigger chance of dying young and being one of evolution's many total losers. Cerebral Canadian rapper Baba Brinkman, creator of the immensely impressive *Rap Guide to Evolution*, even raps about this phenomenon in his remix of Mobb Deep's "Survival of the Fittest" (the video of which embodies male coalitional violence). For these young men, jostling desperately for status and respect, a small insult can explode into violent or even fatal confrontation.

Curtis James Jackson III was just such a young man. He grew up rough in Queens, New York, the son of a teen single mother who dealt cocaine and was murdered when Curtis was 12. After that he lived with his grandparents and sold crack cocaine to make some cash. He was arrested as a teenager for possessing drugs and weapons, and incarcerated for six months. Then in 2000 he narrowly escaped the morgue, shot nine times as he sat in a car outside his grandmother's house.

Before he was shot, Jackson was flourishing as a rapper, having come under the wing of the great hip-hop pioneer, Jam Master Jay. He had already built a considerable following through independent releases, and after his five-month recuperation he recorded his first commercial album, *Get Rich or Die Tryin'*. It made Jackson, or 50 Cent, as he is known to his fans, an instant millionaire and one of the biggest artists of the past decade. Rap is the modern-day art form that best exemplifies the "get rich or die trying" story, but the rise of 50 Cent is a modern-day rendition of a tale told many times over in the history of music. It is the story of Ray Charles, Johnny Cash, Elvis, Frankie Valli and the Four Seasons, The Beatles, Bob Marley and AC/DC. Rock, and now rap, provide in microcosm a view of the competitive, violent, rebellious and desperately showy world that young men have inhabited throughout history. Now and again it is possible for a man, even one who started dirt poor, to achieve the ultimate fantasy of being rich and young at the same time.

Brothers in arms

It isn't always possible to make it on your own. Even 50 Cent has a posse (the G-Unit). Men often need to work together: hunting parties, gangs, military units or rock 'n' roll bands. Men who went to war and lived to tell the tale—be they raiding parties of half a dozen New Guinea highlanders or the Mongol hordes of Genghis Khan—improved their status and thus their evolutionary fitness by vanquishing other groups of men and taking their land, looting their property, raping or capturing their wives and daughters and sometimes even enslaving their families. Throughout our hunter-gatherer past, men often formed tight-knit hunting groups who cooperated to chase and kill animals for meat. And as we saw in chapter 6, the best hunters often snag younger, more fecund and harder-working wives, sometimes landing extra wives and even having more affairs.

The Lamalera people of Lembata, a tiny island in the Indonesian archipelago, traditionally hunt in one of the most unusual and dangerous ways that humans have ever made a living. Teams of men row and sail out to sea in hand-made boats, sneaking up to sperm whales, giant manta rays or sharks. A specialist team member then leaps from the bow to spear the beast with a hand-made harpoon. The sperm whale might just be the most terrifying predator that ever lived, gobbling up giant squid and the even larger colossal squid. Close to a sperm whale, whether in the water or in a small wooden boat, is not somewhere that most people would want to be. But the harpooner definitely has the number-one role in a Lamalera whale-killing team, and harpooners enjoy high status on terra firma too. They turn that status into more children. While men who do not go out in boats have, on average 2.3 children, harpooners and ex-harpooners have about 4.7 children each, and other boat crew members enjoy reproductive success somewhere in between.

Humans are not the only species in which males work in coalitions, but we are probably the most accomplished at it. Sometimes two or three young lions will form a coalition to displace a pride male from his territory, gaining the right to mate with the lionesses on that territory. Bottlenose dolphin bulls form alliances to cut off a female from the pod and forcibly mate with her. In both lions and dolphins, these male bonds are much stronger between brothers or half-brothers, a fact that makes considerable evolutionary sense.

Relatedness eases the evolution of cooperation. It makes more sense to help out a relative than a stranger because we share many of our genes with the relative. So a gene found in one male that disposes him to cooperate with his brother will have a 50 percent chance of being shared by a full brother and a 25 percent chance of being shared by a half-brother. By helping a brother to reproduce, a male can improve the fitness of his own genes. When it comes to cooperation, blood really is thicker than water.

But humans have also mastered another force behind cooperation: we work together because we expect that our efforts will be reciprocated. When they are, the collaboration is strengthened; when they are not, the freeloader is punished or shunned from the team. Our brains are wonderfully equipped to remember not only the faces and names of hundreds of other people but also what we owe them and whether they have done us wrong. These brains make de facto brothers of our coalition mates, be they a hunting party, a sporting team or a platoon. Great coaches and military leaders have long understood this, as did Shakespeare's Henry V in rousing his hopelessly outnumbered troops before thumping the French at Agincourt:

> From this day to the ending of the world,
> But we in it shall be remembered—
> We few, we happy few, we band of brothers;
> For he to-day that sheds his blood with me
> Shall be my brother.

Working together with other men in coalitions was an important strategy in our male ancestors' lives, and as a result it still gives shape to many men's behavior from the locker room to the gang headquarters and, most unfortunately, sometimes even to the office. Rock bands bear more than a passing resemblance to hunting or raiding coalitions. I speculate that they form and stick together via social forces that evolved over eons of hunting and raiding. Somewhere between three and seven musicians, mostly men, pool their talents, energy, aggression and ambition in order to achieve something that no one of them could do on his own. It is also common for bandmates to be brothers.

The Kings of Leon probably set the record for a rock 'n' roll family coalition, comprising three brothers and a cousin. Other noteworthy rock-and-roll brothers include The Beach Boys' Wilsons, AC/DC's Youngs, INXS' Farris', the Finns of Split Enz and Crowded House, Oasis' Gallaghers, Good Charlotte's Maddens, The Everly Brothers, The Allmann Brothers, the three Van Halens, and John and Tom Fogerty of Creedence Clearwater Revival. We should also consider Ann and Nancy Wilson (Heart) as noteworthy rock-and-roll siblings. Fraternity is at the heart of the rock 'n' roll coalition. Consider what rock author Anthony Bozza says about my favorite pair of rock-and-roll brothers, Malcolm and Angus Young:

> The way they hit the strings is AC/DC. They are the electricity, they are the inspiration, they are the devotion to the cause. They absorbed the kind of playing that mattered most to them . . . to create a sound that no one has ever done better.

But even when there are no relatives involved, at the core of most great rock bands exists, even if only for a few years, a brother-from-another-motherhood as strong as Henry V's brotherhood at Agincourt. Jagger and Richards called themselves the Glimmer Twins, feigning not only fraternity but the closest relationship in biology. Lennon and McCartney, too, were like brothers for a time.

But coalitional behavior is only one of the aspects of human nature that evolved as a product of intense male competition. Not every man needs a team to achieve his manly goals of gathering resources and attracting women, and the need for a team waxes and wanes with the circumstances a male finds himself in at the time. Faced with a hostile neighbor, a man will probably seek out the help of those indebted to him. But when a man thinks he can do better by going it alone, he often does. And when one member of a musical coalition decides he can fare better alone, they call it "creative differences." For many members it is like losing a brother, and the magic of the alliance usually remains within the brotherhood. Jagger and Richards were woeful on their own, Lennon and McCartney only slightly less so.

About a girl

> If we look beyond the stage, we discover the sea of impassioned female faces on the other side of the footlights—and the undeniable but often overlooked fact that women's collective participation at the audience level of rock 'n' roll since its inception has been a vital factor in its power as a medium of cultural expression.
>
> Vivien Johnson

Even though rock is mostly made by men, marketed by men, and its history written by men, most rock is made just as much *for* women as for men. The women who listen to the music, buy the records and attend the concerts have been every bit as important in making rock 'n' roll what it was and what it is today.

The machismo of rock is a good match to the faster, louder, power-chord–driven forms of hard rock. Evidence shows that as the machismo in the music and the lyrics amps up, so the number of fans of both sexes falls, but that women drop out of the audience sooner than men. This machismo alone goes some way to explaining why the harder the rock, the fewer women musicians who make it and the fewer women who listen to it. Perhaps a smaller proportion of women musicians have a taste for making hard rock because there is less often a match between the style of hard rock and the things most women musicians want to say.

But the rock that is all about sex and male lust is also, as Camille Paglia says, about women being aroused by male lust. And that is because without women in the audience the music would not be about sex. M is for machismo and metal, but it is also for masturbation. Outside of the hardest forms of rock (and the most gangsta forms of rap), the badness is toned down to a more palatable and often more titillating form. This rock begins to draw its power less from male–male competition and more from female mate choice. In order to appeal to significant numbers of women, music usually needs to be about more than the musician's sexual prowess, the toughness of his homies or anything at all to do with his car.

Some bands achieve this by interspersing hard-rocking "boy songs" with sappy power ballads—ostensibly for the girls. This can sometimes be laughably incongruous. Contrast Aerosmith's saccharine "I Don't Want to Miss a Thing" (incidentally, written by Diane Warren), with the casual misogyny of "Rag Doll" or "Walk This Way" for example. The messages are laughably inconsistent; a band can achieve the feat of speaking in both the voice they use for their girlfriends and the one they use with only their most masculine man-friends around. And they do it without detectable irony and in tight leather pants. Yet despite the layers of crotch-grabbingly hilarious incongruence it can be commercially successful—proving that not everybody listens to lyrics.

It is altogether more difficult to fashion a consistent musical persona in which music and lyrics together appeal to both men and women. According to musician Josh Homme (Queens of the Stone Age), "Rock should be heavy enough for the boys and sweet enough for the girls. That way everyone's happy and it's more of a party." Musicians who accomplish this often do so by articulating complicated emotions or interesting sexual politics. The greatest songs about love are much more insightful than the one-dimensional power ballads of Aerosmith, Guns 'n' Roses or Bon Jovi. Think about The Police's "Every Breath You Take," R.E.M.'s "The One I Love" or U2's "One." They give both men and women thrilling insights into the convoluted minds of the other sex. The bands that do this can achieve the greatest popularity because they connect legitimately with male and female fans. This is what made The Beatles so great.

If there is anything more patronizing than writing separate songs to appeal to men and women, surely it is arguing that women's major contribution to rock is as audience members. I have argued throughout this chapter that we can understand something of the nature of rock by considering how sexual selection operates on men. But the audience is no passive receptacle, and it is important to recognize that the power of this male-driven strand of rock springs from the women in the audience. Vivien Johnson understood this when she suggested that rock and second-wave feminism, despite a lot of open mutual hostility, owe each other a great deal. According to Johnson, the screaming girls and young women typical at mid-1960s Rolling Stones concerts would have experienced "primal collectivity . . . transcendence of the male gaze . . . and the power of their own spontaneous sexuality" in such a way that is "precisely echoed in the primary political objectives of second-wave feminism: female solidarity, autonomy and above all, control of our own bodies."

Early rock won its power by providing the antidote to the sanitized, de-sexed culture in which young women in particular were being raised in the 1940s and 1950s. As he often does, Keith Richards says it best:

> The Fifties chicks being brought up all very jolly hockey sticks, and then somewhere there seemed to be a moment when they just decided they wanted to let themselves go. The opportunity arose for them to do that, and who's going to stop them? It was all dripping with sexual lust, though they didn't know what to do about it. But suddenly you're on the end of it. It's a frenzy. Once it's let out, it's an incredible force.

What might second-wave feminism have achieved were it not for the power young baby-boomer women found in themselves through the Stones and The Beatles?

Although rock derives much of its power from how sexual selection has shaped men, why did the evolutionary forces operating on women not change rock as much? Why have women not made the same inroads into rock that they have into other spheres once dominated by men? First, it is worth repeating that women have made sensational gains across the breadth of popular music. The *Billboard* top ten songs of 2009 included two songs by the Black Eyed Peas (Stacy Ferguson, the most prominent of the four Peas, is a woman), two by the delectably gynandrous Lady Gaga, one each by the women Taylor Swift and Beyoncé Knowles and the men Flo Rida, Jason Mraz and Kanye West, and one by the all-male group All-American Rejects. Women continue to obliterate the old barriers to recording success, but they usually do so in genres other than rock: notably pop and dance. In rock, however, little has changed. The 2009 year-end rock top ten was entirely made up of men.

I have already rejected the idea that boys and men may be more likely to *take up* playing music than girls or women, and the success of women outside of rock reinforces this rejection. I also believe that the success of women within other genres means that men are not necessarily more likely to persist through their tribulations on the long way to the top of music. A culture within rock that is already male dominated and has a long history of often vile sexism—just read any rock memoir from the past 50 years—means women who are drawn to play rock might be less persistent than they are in other genres.

But the hardness and blues-driven nature of rock are musical aesthetics that lend themselves to the things men want to say better than they do for many women, and this skews the music many women make away from rock. Although many women make great rock and enjoy it as profoundly as any man does, this skew means there are relatively fewer of them. In just the same way, there are many more women than men who love Lady Gaga, Sarah McLachlan or Natalie Merchant.

While a largely male agenda shaped the musical boundaries we now recognize as rock, women blazed new trails by finding new female sexual personae to express and then legitimizing them. One need only look at women like Madonna in the 1980s or Beyoncé today to see raunchiness expressed in a credibly female persona. Mainstream rock success by women like P!nk will probably lead to further success, although we should not pin our hopes too high. Thirty years ago, Deborah Harry, the singer from Blondie and one of rock's most visible women, acknowledged the role of sex in rock, and pointed to a future with many more original female personae: "Sex sells and I do exploit my sexuality . . . women are going to be the new Elvises. That's the only place for rock 'n' roll to go. The only people who can express anything new in rock are girls, and gays." I think it would be fair to say that we are still waiting for our first female Elvis. Although personally I'd be just as intrigued by a long-lived anti-heroine to offset Keith Richards.

11
Immortality

The music business is a cruel and shallow money trench, a long plastic hallway where thieves and pimps run free, and good men die like dogs. There's also a negative side.

—ATTRIBUTED TO HUNTER S. THOMPSON, 1988

Rock stars seldom grow old gracefully. They tend to die young, fade away, or end up shattered, broken and on reality television. Dead and aging rock stars reveal much about why some people die young, why others live well into old age, and why our minds and bodies tend to deteriorate as we get old.

KURT COBAIN WAS THE MONUMENTAL talent who fashioned the defining sound of my generation; the messiah who saved rock after the mediocrity of 1980s cock rock and hair metal. But behind his public eminence stalked a personal hell of addiction and depression. He blew his head off with a shotgun at the age of 27, leaving a note quoting Neil Young's "My My, Hey Hey (Out of the Blue)": "It's better to burn out, than to fade away." For a rock star, dying young or fading away does seem to be the brutal choice. But that is really true for everybody. This is a chapter about burning out and fading away; dying and aging. The fast-living rock 'n' roll lifestyle accelerates both, but everybody else either dies young or dies slowly. Rock is really just life, amplified, and turned up to 11.

Those who survived rock 'n' roll are now feeling their age. Chuck Berry is still charging hard at 84, Little Richard is 78, and McCartney, Dylan and the surviving Stones are nearing 70. The baby boomers who listened to them when they were young are aging too. Politicians now obsess about how their baby-boomer cohorts will strain healthcare and social security as they retire and their health deteriorates. The massive number of babies born in the 1950s and 1960s also survived better than any generation before them, reaping the benefits of immunization programs, hygiene improvements, plentiful food and a low rate of death from violence and warfare. Western societies include an unprecedented number of people over 65, many of whom are now encountering the diseases and afflictions that characterize old age.

As a result of the wonderful developments that have brought down the death rate for children, young adults and the middle-aged, the commonest causes of death today are very different from 100 years ago. In 1907, 20 percent of Australians died from circulatory diseases and only 8 percent from cancer. By the year 2000 those numbers had risen to 39 percent and 28 percent respectively. As the number of people living to older ages climbs, other diseases of old age will become our biggest killers. Dementia afflicts half of Americans over 85, and Alzheimer's disease is the biggest cause. Can you imagine the suffering likely to unfold, and the right-to-die controversies that will probably rage as a generation succumbs to dementia?

Death is one of life's few certainties, and if you live long enough then aging appears inevitable. By aging, I don't simply mean growing older; I mean the way our bodies deteriorate as the years go by so that with each passing year we are more likely to die than we were in the previous year. It seems childish to ask "why do we die?" And "why, if we are lucky enough to live a long time, is dying preceded by a period of deteriorating old-age?" But the best questions are often asked by children.

In this chapter I explore the evolutionary reasons for aging and dying. Understanding more about how aging evolves will be essential if the ever-longer lives we lead are to be happy and healthy ones.

Imagine all the people

Two fallacies envelop the questions of death and aging: that we somehow need to die and that falling apart with age is inevitable. It is tempting to think that in death we do our adaptive part for our children and grandchildren by getting out of their way and not using up valuable food, space and oxygen. This idea says more about how many people view the old than it does about why we age and die.

Imagine a world in which everybody who reached their allotted sum of years shuffled magnanimously off the perch to make way for the next generation. It might sound harmonious, but it would also be too easy to cheat. Individuals who outstayed their welcome for a few extra years would lose nothing and potentially gain a lot. They could help rear their grandchildren, for example, allowing their sons and daughters to have larger families. A gene that extended life would, within a few generations, overrun the genes that cause folks to kindly expire when their meter runs out. The costs of supporting long-lived mutants are externalities shared by everybody, and yet the benefits of increased fitness go to the mutants alone. Growing old and dying are not adaptive. As we shall see, however, they are an incidental and unfortunate consequence of natural selection.

Not only do we not need to die, but we are really all immortal. At least our ancestors' genes have proved immortal so far. Everybody alive today is the descendant of hundreds of thousands of generations of successful ancestors, going back to the very first life forms on earth. Those first ancestors passed their genetic information on to their offspring, the offspring passed it to the grand-offspring and so forth. With every generation a few small mistakes—mutations—arose in that information.

Some mutations tweaked the way proteins worked together to make cells and, eventually, bodies. Most tweaks were catastrophic and the young who inherited those mutations died before they could pass them on. But a few tweaks improved how the offspring reproduced and passed on the tweaked information.

Every individual living today has inherited the very first DNA molecule, but that molecule has been copied, miscopied, expanded and changed by the cycle of mutation and natural selection over billions of years. Now every one of us carries a DNA sequence that is unique, and that is much different from the first DNA. The story of evolution is the story of that information. How that information makes bodies and behaviors is just a very interesting subplot.

Even after 4 billion years of relentlessly being copied and passed from parents to offspring, our genetic information hasn't fallen apart with wear and tear. The many mistakes that have occurred have either been corrected by the DNA's own repair mechanisms or they have been lost or preserved by natural selection. The DNA in our cells today, and in the cells of all the microorganisms, plants and animals with which we share this world, has thrived, uninterrupted, partly by fixing mistakes, partly because the information has been phenomenally successful at making copies of itself, and partly by good luck. But even though our genetic information persisted stubbornly through 4 billion years of life on earth, bodies are not so hardy.

Unlike genetic information, bodies are temporary. It is bodies that show wear and tear but even this is not inevitable. Some sea anemones and their cousins—such as the tiny translucent *Hydra*—reproduce simply by budding off part of their bodies. They have no specialized set of organs for reproduction. Instead, the adult "is really a perennial embryo," according to Daniel Martinez, the world's foremost authority on aging in *Hydra*. According to Martinez, "the genes that regulate development are constantly on, so they are constantly rejuvenating the body." As a result, *Hydra* don't seem to age at all. Other animals as small as *Hydra*—less than an inch long—typically last no longer than a few weeks before deteriorating and dying. But when Martinez studied a large number of *Hydra* in the lab for four years, very few died and they showed none of the symptoms of aging: the death rate did not increase and the rate at which they budded off new polyps didn't slow.

Hydra evolved the ability to rejuvenate themselves because that makes them more effective at reproducing. Chop off part of a *Hydra* polyp and it will grow into another fully formed *Hydra*. A polyp is always ready to grow another polyp out of the side of its body and reproduce. It makes sense, then, for the whole body to be kept shipshape and ready for reproduction. For humans and most of the animals we are familiar with, producing a new life is far more elaborate than budding off a perfect clone. We have evolved specialized organs for reproduction, and within those organs there are special cells that make the eggs and sperm. Only the genetic information in those cells takes the next step toward immortality, inherited by the new offspring. It is no surprise, then, that the body maintains these cells obsessively. They alone are the future.

Because bodies can be killed by predators, parasites and diseases or die in accidents, even the best-maintained bodies don't live forever. Making bodies that last makes little sense if the investment in durability detracts from the main job of issuing new copies of the genetic information. Unlike *Hydra*, for most animals a body is merely the disposable hardware that the information makes and then uses to spread itself. How long an individual can expect to live and how fast it deteriorates in old age depend very much on how disposable bodies are.

More than just reasonable wear and tear

My eight-year-old Subaru is showing signs of aging. I recently spent one-fifth of the car's current value replacing the clutch, and there are faulty window motors and light fittings that nag at me. Each year I spend more money repairing and maintaining it than I did the year before. But I am inclined to accept that things just aren't built to last. At least I was until I learned that some Boeing 747 airplanes in service are older than me. If planes can last 40 years without falling out of the sky, then why don't cars keep running smoothly for decades?

Commercial airliners cost millions of dollars and even minor safety incidents can be bad for business. So airlines maintain their planes meticulously. Cars are a lot less expensive and most breakdowns less catastrophic, so even though we maintain our cars we don't do it with the obsessive diligence of a Qantas maintenance crew. Both planes and cars can be kept in superb working order by investing in the best care and maintenance, but wear and tear accumulates more on cars because the benefits of maintenance and the costs of things going wrong are not as great as they are for airliners.

Just like cars, our bodies suffer much wear and damage that accumulate with age. It is tempting to think of aging as the inevitable accumulation of a thousand tiny insults to our bodies, but just as teams of dedicated engineers maintain airliners, our bodies employ their own defenses to prevent and repair damage. Our cells deploy hundreds of molecular tools to fight infections, expel foreign objects, soak up free radicals, neutralize toxins, heal wounds, and even correct copying errors in our DNA. Experiments on long-lived and short-lived animals show that some of the biggest differences in life span come about because there are differences in how hard the repair mechanisms of different animal species work. Long-lived animals tend to have better repair mechanisms whereas shorter-lived animals are less well equipped for repair.

The repair mechanisms in animal bodies evolve to suit the way those animals live their lives. But why don't they work at maximum capacity all the time? Just as it would be unnecessary and unprofitable for an airline to overhaul every jet after each flight, it would be just too costly for bodies to deploy their full repair tool kit all the time. And that is part of the key to why we age: our repair mechanisms work hardest in young people, but as we get old it seems that many of these repair mechanisms work less efficiently. It's very much like the way we keep our cars well maintained when they are new, but our conscientiousness wanes with the car's resale value. Over evolutionary time, genes that kept young individuals in top shape were favored more strongly by natural selection than genes that kept old bodies from falling apart. But why?

Selection gets weaker with age

Forget about my car. I really love my iPhone. Make no mistake I am not an irritatingly smug Apple-ite, but my iPhone has just the right blend of simple elegance, clever functionality and smart applications for me. All the same, I don't expect this phone to last. To my great irritation, my mobile phones have a habit of breaking before it is time to renew my contract. Why can't a phone last longer than the contract I bought it on without it getting so much dust inside it that the screen becomes impossible to view, and without the battery life getting so short that the whole device becomes entirely un-mobile, tethered to a charger like a patient on life support?

The short answer is that phones break because of wear and tear, but this only begs the question. Why aren't phones engineered to last five years of use by a normal person (that is, me)? To partly answer this, let us consider what might happen to a consignment of 100,000 mobile phones between assembly line and the bottom drawer where once-loved gadgets go to die. A small number of faulty phones never make it out of the factory. Others are lost in accidents or are simply misplaced somewhere along the supply chain. These problems together might prevent 5 percent of phones from ever being sold. That is a hefty 5000 phones, and you can see why the manufacturer would want to shave down the percentage of these phones that are never sold.

But a faulty phone that needs to be repaired or replaced within the two-year warranty period is just as costly to the manufacturer, so why don't they care as much about fixing problems that occur after a year or so of solid use as they do about avoiding accidents and inefficiencies in their distribution? Let's look at the remaining phones. Some are lost or stolen. Close friends and I have lost phones by accidentally dropping them over cliffs, under car tires or in the toilet and by leaving one in a pocket during an unscheduled swim. Other phones owned by technophiles with short attention spans get replaced and forgotten before the warranty expires. After all these mishaps, there might be 30,000 phones in the hands of people who would like them to keep on working for longer than two years. A problem that occurs in 5 percent of phones that make it this far will only result in 1500 phones breaking and needing to be repaired or replaced, which is a lot less of a problem for the manufacturer than the 5000 phones affected by a similar problem in new phones.

Of course this is a silly fictitious example, but it is designed to illustrate something important about bodies. A problem that causes 1 in 20 phones to stop working after two years is nowhere near as costly to the company as a problem that happens in 1 of every 20 new phones because so many phones are lost, stolen, broken or simply discarded before they ever get to two years of age. In exactly the same way, imagine a form of cancer that causes 1 percent of all people to die before their 20th birthday. This disease would kill far more people than another cancer that never develops in anybody less than 50 years old, yet kills 1 percent of those who make it to 50.

Now imagine that all the people who get the young people's cancer do so because of a single genetic difference from the rest of humanity. Similarly the old people's cancer is due to a different yet equally simple genetic mutation. What would happen under natural selection? In the case of the young people's cancer, if every carrier died before the age of 20 then very few carriers of the mutant gene would get to have children, and those who did might only have one child before falling ill. As a result there would be exceedingly few babies in the next generation carrying the disease. Natural selection is relentlessly efficient at eliminating genes that diminish lifetime reproductive success—especially genes that kill children before they become adults. This is why so many genetic diseases of childhood and adolescence are due to exceptionally rare genes—because natural selection usually all but eliminates this kind of gene.

A late-acting cancer gene, on the other hand, has a much easier time. By 50, most parents have finished having babies and many have even fledged their children out into the wide world. Dying after 50 has a far less dramatic effect on a parent's evolutionary fitness than dying at 20 would. It is possible that those men who die after 50 would lose out on the chance to sire one or two more babies, and both moms and dads miss out on a chance to help their own children establish themselves in the world and raise a family of their own. But neither of these costs to fitness are anything near as dramatic as dying before you have ever had a chance to reproduce.

Not only do 50-year-olds have far less of their reproductive future ahead of them than 20-year-olds, but many people die between the

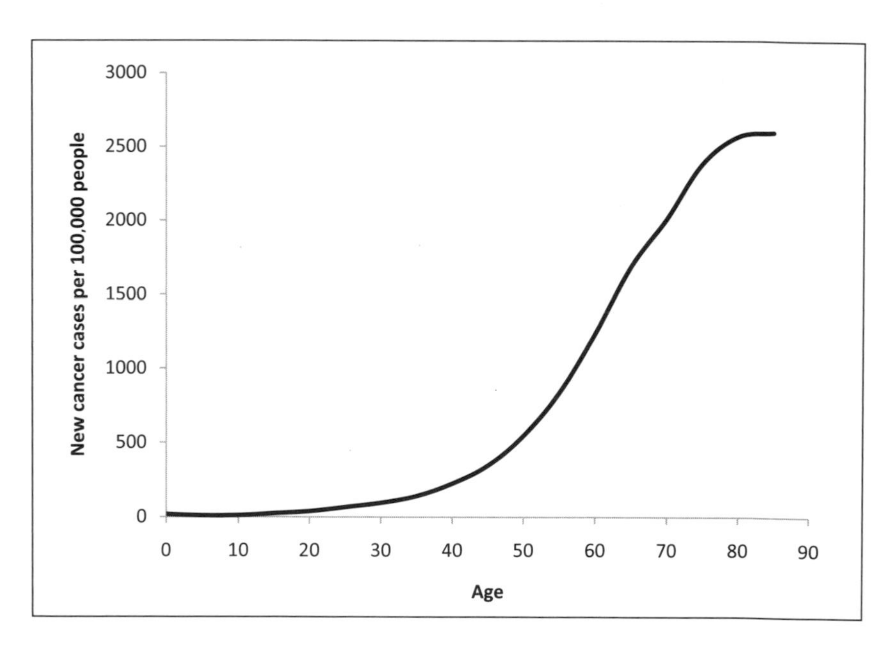

The risk of being diagnosed with cancer increases steadily throughout life. Figures based on Australian data for 2005.

ages of 20 and 50 of causes unrelated to the cancer. So natural selection against the over-50s cancer gene can never be as powerful or as efficient as it is against a young person's cancer gene. The fact that our bodies deteriorate with age can be pinned on this very simple fact that the strength of natural selection declines with age.

The very real chance of being diagnosed with cancer at different ages illustrates my point (see figure on p. 256). In Australia fewer than 20 out of every 100,000 children under the age of 15 develop any kind of cancer each year, but the incidence of cancer grows steadily thereafter so that more than 1 percent of Australians over 60 years old are diagnosed with a new cancer each year. Adults over 60 are 500 times more likely and those over 70 are 1000 times more likely to develop cancer in the coming year than a child is.

The mechanisms that prevent and repair DNA damage that can lead to cancer, and that fight and suppress the rogue cells that can become cancers, are hundreds of times more effective in young people than in old people. Natural selection has fine-tuned those mechanisms because young people with faulty defenses against cancer tend to die before they ever reproduce. Older people with genes that make their cancer defenses faulty are often parents, and sometimes even grandparents, before they find out they have cancer.

Natural selection efficiently prolongs life span by eliminating genetic diseases, keeping the body running smoothly and controlling cancers until the age when we tend to have babies. It becomes ever less efficient with each subsequent year. So the frailties that cause us to look, feel and act ever more decrepit do not get eliminated or repaired by natural selection as effectively as similar problems would in a 15-year-old. Beyond the ages at which our ancestors reproduced or—as grandparents—helped their own offspring to reproduce, selection has always been very weak. This is the valley of the shadow of natural selection, where evolution has been powerless to ameliorate the ravages of old age.

Thrive now, age later

It isn't only that natural selection is much weaker on genes that cut us down in old age; some of those genes have a positive effect on fitness early in life. Aging is an unfortunate side effect. These genes can be quite common because the advantages they bestow early in life make them especially likely to be passed on. One such gene, called *APOE*, produces a protein called Apolipoprotein E that has the routine but essential job of breaking down a variety of complex molecules that accumulate in the body. There are more than 30 forms of the gene of which three are reasonably common: ApoE-ε3, which is considered the "normal" form and is most common, as well as ApoE-ε2 and ApoE-ε4. If proteins were sentences and the amino acids from which genes assemble protein were the words, then the Apolipoprotein is 299 words long. That is long even for one of my sentences, but 299 amino acids is short for a protein. The proteins made by the ε4 (epsilon 4) and ε2 forms of the gene contain one mistake each. In each case one of the 299 amino acid "words" is wrong, a small mistake with big consequences. It is the mistake in the ε4 form of the gene that is of greatest interest to my story.

Epsilon 4 is implicated in atherosclerosis, the thickening of artery walls due to the build-up of cholesterol and other fatty molecules, and in Alzheimer's disease, the incurable degeneration of the brain that is the most widespread cause of senile dementia. Alzheimer's is a classic aging disease: it tends mostly to manifest after the age of 65, well after parents have had a chance to raise their children, and often their grand children too. It is among the cruelest of afflictions, robbing sufferers first of the ability to form new memories, and then progressively of all their existing memories, sense of self and eventually their dignity. It is the disease that claimed the life of former US President Ronald Reagan, actor Charlton Heston, and the philosopher and writer Iris Murdoch. One of my favorite authors, Terry Pratchett, has an early-onset form of Alzheimer's, and his battle with the disease is the subject of a number of documentaries. In 2010 Alzheimer's claimed the life of George C. Williams, the genial intellectual giant who did more than anyone else to show why natural selection doesn't eliminate genes like ε4.

Around one in four people have one copy of the ε4 version of the gene and one copy of another form, and these people are about five times as likely to develop Alzheimer's as people without any copies of epsilon 4. Around 1 in 50 people have two copies of ε4, and these people are around 20 times more likely to suffer from Alzheimer's than normal. With such chilling risk factors it is little wonder that people who have had their whole genome sequenced, including James Watson, co-discoverer of the structure of DNA, and psychologist Stephen Pinker, have elected not to learn which versions of the *APOE* genes they carry. Until effective tools to remediate the effects of epsilon 4 are available, many people would prefer not to know. Genome pioneer Craig Venter, on the other hand, publicly stated that he has one copy of epsilon 4. He is already taking cholesterol-reducing drugs as a preventive measure against atherosclerosis, and if cholesterol is part of the link between epsilon 4 and Alzheimer's, he might foil dementia too.

Difficult as it is to imagine, there is an upside to epsilon 4. Several recent studies have shown that young adults with one or two copies of ε4 do better on intelligence tests and in tests of memory and attention span. A study in the Czech Republic showed that ε4 carriers are less likely to drop out of school and much more likely to go on to university than non-carriers. Epsilon 4 might actually be beneficial early in life, giving carriers an advantage in memory, intelligence and the ability to focus on important information. Even though the late-acting effects of epsilon 4 are both devastating and terminal, the selection on the 65+ age group in which Alzheimer's most often develops is weak. Selection on young adults, by contrast, is very strong, because few have died yet and they have the majority of their reproductive career ahead of them. While there is still much debate about the relative importance of various types of selection in shaping intelligence, it has always been profoundly useful to be smart. So even a small effect on intelligence and memory can offset the enormous effects that ε4 has on carriers late in life.

Most genes are, like *APOE*, involved in several biochemical processes in different parts of the body. It makes little sense to view the action of a gene in terms of one function that it performs or a single problem that a mutant form of the gene causes. The effects that a gene has early in life can often outweigh, in fitness terms, the costs that the gene imposes late in life after reproduction has usually finished and the kids have moved out of the home. As a consequence, genetic variants like ε4 are disturbingly common in the population, much more so than any rare late-acting genes that we might in future eliminate via genetic screening.

Death is a side effect of reproduction

The connections between ε4, early intelligence and both atherosclerosis and Alzheimer's late in life are subtle, but they illustrate a straightforward evolutionary scenario. A gene that improves fitness early in life is favored by natural selection, but that advantage comes at a terrible price late in life. Our bodies make hundreds of these kinds of trades, investing in reproduction in early adulthood at the likely expense of reproduction and longevity later on. Some of these trade-offs come as a consequence of our actions.

Nothing is free, not even love, and especially not reproduction. At every step, reproduction exacts its heavy price. Consider the costs to the female cane toad who makes 100,000 eggs at a time; the male elephant seal who fights off all comers to hold a beach; the mouse who doubles in weight as she carries a litter of eight or more pups to term; the elephant cow who spends 20 years teaching her daughter to be a responsible grown-up elephant. These costs can include an immediate increase in the chance of death; childbirth complications killed a big fraction of adult women for most of human history, and still do in many parts of the world.

The costs can also accelerate aging because reproduction requires energy, protein and other resources that would otherwise be used for maintaining and repairing the body. When an airline cuts maintenance costs to increase profitability, planes accumulate wear and tear. As a result, the planes don't last as long before they have to be replaced, and sometimes they fail with tragic consequences. In the same way, each time a body diverts resources from maintenance to reproduction, some damage might go unrepaired. The result is aging and, sometimes, an early death.

Among British aristocrats born between the years 740 and 1876, the younger a woman was when she first gave birth, the less likely she was to live beyond 60. Women who lived beyond their 80th birthday were more likely to be childless, and those octogenerians who were mothers had had fewer children than those who died younger. The effects of childbearing on women's life span can still be seen today. My colleague Alexei Maklakov gathered data from 205 modern-day countries and territories, where the average number of children a woman bears in her lifetime ranges from 0.9 (Hong Kong and Macau) to 7.1 (Niger and Guinea-Bissau). Women live longer in countries where they have fewer babies, even when Alexei controlled statistically for other variables like average income and population density. Those of us living in industrialized societies are used to women outliving men by several years, but in countries like Afghanistan, Niger and the Federated States of Micronesia, where fertility is high, men usually live longer than women.

To our bodies, the future is seldom worth as much as the present, and so we discount the future in making our reproductive decisions. But just how bright the future looks can often depend on an individual's circumstances. It pays to be sensitive to changing conditions, trading more of our future away when prospects are poor but keeping more in reserve when we can expect to live long and prosper. According to anthropologist and psychologist Daniel Nettle, exactly this seems to be happening in contemporary England, where people in poorer neighborhoods die slightly younger but have 25 fewer years of good health than people in wealthier neighborhoods. In the poorer neighborhoods, women have their first baby at an average age of 22, whereas first-time moms in more affluent parts are, on average, older than 28. Poorer women also have smaller babies and they breastfeed for less than two months, whereas wealthier women breastfeed for longer than four months on average. It seems that women who expect to have fewer years of good health start reproducing earlier and invest less in each child.

It remains to be seen whether early childbearing contributes in turn to poorer health, creating a vicious cycle driven by poverty. A recent study from Germany suggests it might. In the aftermath of World War II, when the Allies partitioned Germany into the communist German Democratic Republic (East Germany) and the Federal Republic of Germany (West Germany), they inadvertently created an interesting natural experiment. Policies in the East encouraged mothers to reproduce, supported their return to work, and made single motherhood much easier than it was in the West. The tax and social security system in West Germany promoted "male-breadwinner and female-housekeeper" roles. As a result, fewer East German women were childless, mothers started having children younger, and many more mothers worked during their child-rearing years than in the West. The consequences of these differences could still be felt in 2006, 16 years after re-unification. Mothers and fathers who raised four or more children in West Germany were more likely to be in good health than those who had only two children. But East German mothers were in much poorer health, and the more children they had, the poorer their health and the shorter they lived.

The price of attraction

The costs of carrying a baby, giving birth and then breastfeeding are palpable. The costs of finding and attracting a mate and vanquishing competitors are a little harder to observe directly, but they are every bit as real. Animal courtship displays often greatly increase the show-off's chances of dying. Male frogs that call all night in the desperate hope of attracting a female also put themselves in mortal peril of being picked off by an owl, a snake or a bat. Even if they survive the night, the exertion of calling at well over 100 decibels can leave them shattered and spent. Male long-tailed widow birds flap pathetically across the African veld, barely able to keep airborne as their tails—four times their body length—drag them down. But females notice these efforts and choose to nest near and mate with the longest-tailed males. Sexual selection is so potent that it exaggerates sexual signals until the costs become overwhelming.

Sometimes, slow and steady is the best way to go. Display too much and you'll soon be dead, unable to recover from too much calling or to fly far enough to find food. Some individuals grow only a modest tail or call for only a few hours a night, in the hope of being able to come back and do it again. But every day that an animal lives brings it one day closer to dying. So older individuals sometimes put more effort into being attractive than they did when they were younger, with less to lose if they overcommit.

But that isn't always true. When life is likely to be short, it can pay to burn bright, even if that means the brief candle flickers and occasionally extinguishes too soon. In a study led by John Hunt, my research group showed that female field crickets fed a good diet tend both to make more eggs and to live longer than females on a poor diet. But give a male a good diet and he will call so relentlessly that he burns himself out and dies long before males fed the poorer food, who soberly call for a few minutes a night.

Crickets given the good diet live by the same motto as Pretty Boy Romano, the young thug in the 1949 movie *Knock on Any Door*, who famously said, "I wanna live fast, die young and leave a good-looking corpse." Young men take far more crazy risks than women and as a result are much more likely to die or be critically injured in accidents or altercations. Men are also four to five times as likely to die from suicide and have a higher risk of dying from parasitic and infectious diseases. In other mammals the elevated risk of death from infectious disease is associated with the intensity with which males compete for access to females. The link between disease and competition comes courtesy of the male hormone testosterone, which controls the timing and extent of male aggression and also tends to suppress the immune system. Testosterone has many nasty side effects beyond all the crazy risk-taking, aggression, suicide and increased risk of disease. It makes men more likely than women to suffer from heart disease, arteriosclerosis and other aging diseases.

These masculine maladies are not equally strong in all places and all the time. Dan Kruger and Randolph Nesse at the University of Michigan studied data from 14 Eastern European nations before (1985–89), during (1990–94) and after (1995–99) the transition from centrally planned to market economies. The transition created new economic opportunities for those enterprising and driven enough to take them, and as a result inequality in income and status skyrocketed. This kind of change is exactly where we would expect people to take big risks, overwork themselves and suffer astronomic stress. And evolutionary theory predicts that men will suffer far more in this way than women. And that is exactly what happened. Men died sooner, relative to women, during and after the transition than they had before the end of the Cold War. But in Western Europe, there was no such change in patterns of mortality over the same periods.

Into the black

Which brings us back to rock 'n' roll. In the previous chapter I wrote of the reckless bravado that runs through rock and much of popular music; the relentless striving for success, fame, fortune and gratuitous sex with groupies. It isn't the only theme, but it is an important one and it embodies an equally important evolutionary principle: that the future isn't worth as much as the present. Or as Jim Morrison put it, "I don't know what's gonna happen, man, but I wanna have my kicks before the whole shithouse goes up in flames."

Throughout history, poor men from the wrong side of the tracks have been especially inclined to discount the future and to risk everything they have on the small chance of a big evolutionary payoff; men who seethed with rage, ambition and creative fire—like blues pioneer Robert Johnson (1911–38). According to music journalist Stephen Davis,

> In the delta of the Mississippi River, where Robert Johnson was born, they said that if an aspiring bluesman waited by the side of a deserted country crossroads in the dark of a moonless night, then Satan himself might come and tune his guitar, sealing a pact for the bluesman's soul and guaranteeing a lifetime of easy money, women, and fame. They said that Robert Johnson must have waited by the crossroads and gotten his guitar fine-tuned.

Whether acquired by Faustian deal or old-fashioned hard work, Johnson's guitar playing and innovation guaranteed him immortality. Yet he was rendered mortal by a husband's jealousy, a bottle of booze and some strychnine at the unlucky age of 27. Just as the sexually selected plumage, calls and pheromones used by animals to attract their mates—and the weapons and aggression that males use to secure territories and mates—are costly, so musicians pay a heavy price for their success.

The frantic striving for stardom, status and success that defines younger rockers, rappers, blues musicians and many other entertainers requires that they discount the future. They work a hard day's night, live on the edge and many are prepared only to sleep when they're dead. They

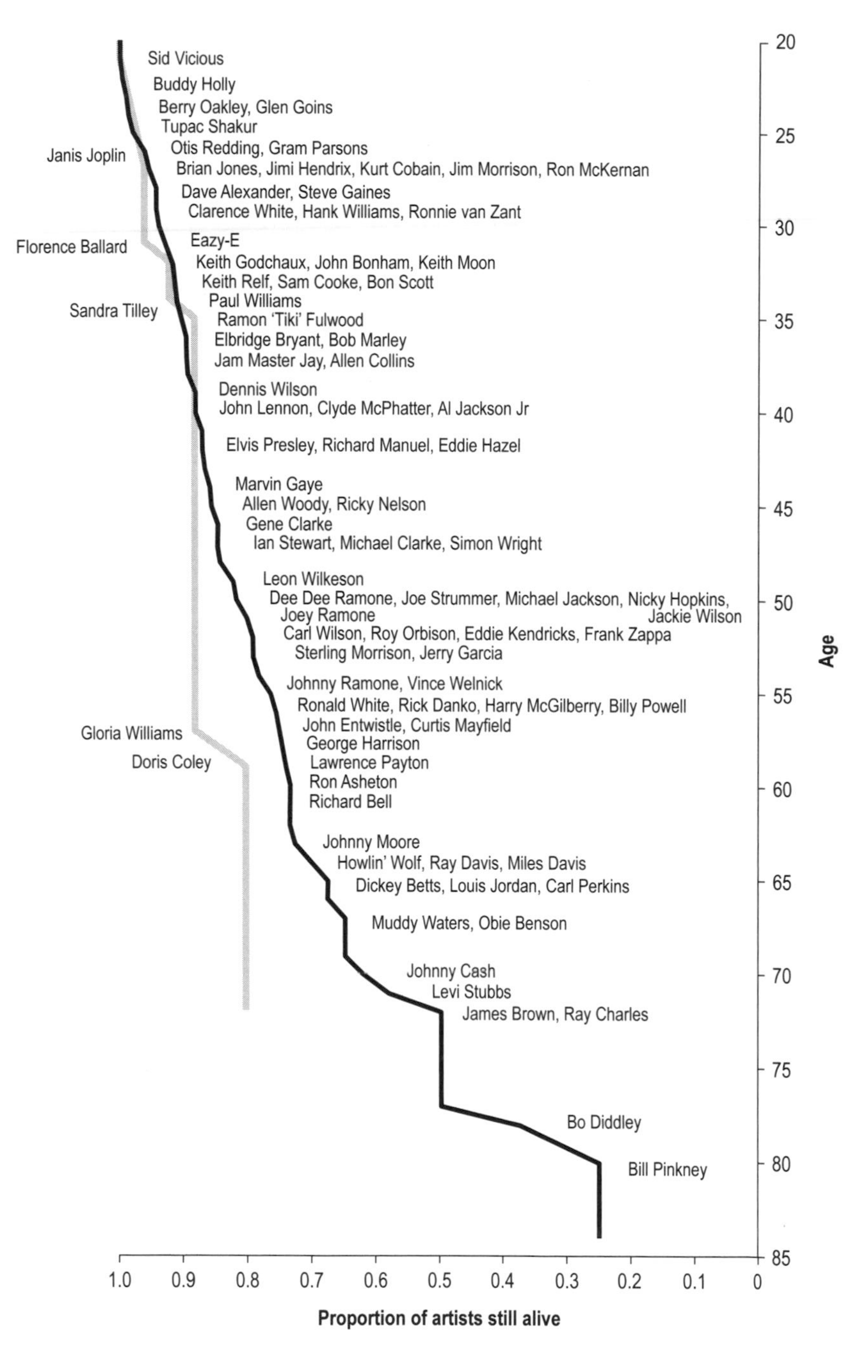

Mortality of *Rolling Stone* Immortals. The proportion of women (grey) and men (black) still alive in relation to age. Position of names indicate age at death.

also use drugs, abuse booze, drive their cars too fast, carry weapons and escalate deadly rivalries more than average people of the same age.

Robert Johnson, Kurt Cobain and Brian Jones were by no means the only greats to die in what should have been their prime. In fact Jimi Hendrix, Janis Joplin, Jim Morrison, the Grateful Dead's Ron "Pigpen" McKernan and, most recently, Amy Winehouse all died at 27. And there is nothing special about 27; the twenties are a perilous time for rock stars and things don't get much better after that (see graph on p. 266).

Mark Bellis at Liverpool John Moores University and his colleagues recently showed that between 3 and 25 years after first becoming famous, performers from the all-time Top 1000 albums in rock and pop music are more than 1.7 times as likely to die as North Americans or Europeans of the same age. They died from problems related to drugs and alcohol (31 percent), accidents (16 percent), violence and suicide (9 percent), cardiovascular disease (14 percent) and cancer (20 percent). If popular music like rock, rap, jazz and the blues is the greatest courtship signal that ever evolved, it is also one of the most deadly.

If rock doesn't kill you it makes you older. Apart from possibly Keith Richards, nobody better personifies the devastation wrought on a body by the rock 'n' roll lifestyle than the Prince of Darkness. Ozzy Osbourne considers himself a medical miracle; he spent 40 years on a massive drink-and-drugs bender, broke his neck in a quad-biking accident and has twice died and been revived. He also escaped unscathed when a light plane carrying his guitarist, Randy Rhoads, tried to buzz Osbourne's tour bus in a fit of rock 'n' roll stupidity and collided with it instead—killing all three occupants of the plane. Any viewer of his reality television show can tell that Osbourne has lost some of his hearing, suffers from tremors and appears visibly damaged from his four decades of hard living. Ozzy is a rock survivor, but he carries the scars.

Intriguingly, Ozzy's genome was recently sequenced, ostensibly to find out why he's still alive after all he's been through. It is also an ingenious public-relations coup for private genomics and the company involved. According to Osbourne, the results say:

> I'm 6.13 times more likely than the average person to have alcohol dependency or alcohol cravings (er . . . yeah); 1.31 times more likely to have a cocaine addiction (this must be bollocks, because anyone who takes coke as much as I did gets hooked); and 2.6 times more likely to have hallucinations while taking cannabis (makes sense, although I was usually loaded on so many different things at the same time, it was hard to know what was doing what).

Whether genomics is the most fruitful tool to understand how Ozzy became a rock survivor rather than a rock statistic remains to be seen, but he has certainly defied the biological odds.

Robert Johnson, Brian Jones, Jimi Hendrix, Janis Joplin, Jim Morrison and Kurt Cobain all died at 27 but their achievements give them a kind of immortality that many lust after yet few people ever achieve. They are forever preserved in our memories at their prime, all youthful looks and tragically creative fire. No matter how many Hendrix, Doors or Nirvana T-shirts are worn by generations that were unborn when the superstars drew their last breath, they will never lose their luster.

These artists will never have to shimmy and pose on the stadium screen, approaching 70, Jagger-like, playing songs and busting moves from a lifetime ago. Or bear the indignity of drug-fucked reality television parody, recycled to feed the celebrity magazine pulp mill. Their fire won't dim and their talent won't atrophy. And they will never surrender, unable to play or sing, as they give in to arthritis, Parkinson's, cancer and Alzheimer's. But glorious as that kind of immortality is, it is no use to Robert, Brian, Jimi, Janis, Jim or Kurt. After all, as Neil Young emphasizes when he sings "My My, Hey Hey (Out of the Blue)," "once you're gone, you can never come back."

Thanks

I REALLY DO HAVE A dream job. I get to think about the most important idea anyone ever had and try to answer some of the most interesting questions people have ever asked. And I get to think about sex. It is an enormous privilege and pleasure to spend my working life teaching, learning and discovering new things. It is an even greater privilege to work in a field where we do things not because they are commercially lucrative, or because they will save the world, but rather purely because they are interesting. Most of what we know about evolution was discovered by scientists studying what interests them, and I hope this book has given you a taste of just how fascinating evolutionary biology can be. I hope I have also opened the curtains a little on how important an evolutionary worldview can be.

Not every society feels it can afford to support scientists who study interesting questions. The Australian Research Council has generously supported my research since 1999, and for the last six years has funded part of my salary from its fellowship schemes. This gives me both the time and the freedom to pursue the research I find most interesting. Funding for curiosity-driven research is essential if a society is to have any kind of intellectual pulse, but basic science is woefully underfunded in most nations. I hope that this book provides a small taste of the benefits of publicly funded science.

I have been extremely lucky to work at the University of New South

Wales for a decade, among supportive, interesting and friendly colleagues in the Evolution & Ecology Research Centre and the School of Biological, Earth and Environmental Sciences, many of whom have helped me pound my ill-formed ideas into somewhat more coherent chapters. My own research group have been enormously patient with me, allowing me to road-test the ideas in this book over coffees and lab meetings. They have also patiently understood this detour into popular writing, even when it meant diverting time and effort away from many of our collaborative projects and manuscripts.

Several friends, some of whom are my research students, and some of whom are my colleagues, listened to far too many monologues about this book and read many, or even all, of the chapters. Without their perspicacious advice, kind encouragement and willingness to call "bullshit" from time to time, this book would never have been finished. They include Emma Johnston, Sam Maresh, Alex Jordan, Edith Aloise King, Russell Bonduriansky, Ray Blick, Margo Adler, Mike Jennions, Michael Kasumovic, Beth Kasumovic, Carla Avolio, Simon Griffith, Matt Hall, Genevieve Quigley, Bill von Hippel, Brendan Zietsch, Elke Venstra, Gemma Smart, Juliette Shelly and Aileen Lee. Jules Shelly also created a gorgeous website and built our blog on evolution in the modern world <www.robbrooks.net>. I am indebted to Lyn Craig, from UNSW's Social Policy Research Centre, who took the time to discuss the contributions that men and women make to households, although the paper she sent me did not make it into the final version of this book.

My former research assistant, Nicolle Spyrou, gathered all kinds of crazy data for me, including biographic information on the *Rolling Stone* Immortals and details on 55 years of *Billboard* number one records. She also kept the lab running smoothly and very happily for more than seven years. I thank her, and her wonderful successors, Elke Venstra and Heather Try.

This book began with a call from Stephen Pincock, then a publisher at NewSouth Books, who talked me into the project, helped get it started and gave it shape. Although he left the press when this book was half-done, Jane McCredie replaced him with aplomb. She prodded me to make important changes where necessary, but showed a calm belief in my writing, which gave me the confidence I needed. Without Stephen's encouragement I would never have started. Without Jane's confidence I might not have finished. I was a very nervous first-time book author, yet Tim Fullerton not only edited the manuscript with great patience and attention, and made some wonderful suggestions that I worked into the final manuscript, but also constantly reassured me that we were on the right track.

My parents, Ben and Patti, taught me to read when I was young and taught me to *write* in my last year of high school. Their encouragement to follow my interests eventually led me here. Mid-book, when Mum was ill, they housed me, Jacqui and the kids for two months and made the space for me to maintain some momentum when really it was them who could have used my help. They also edited and commented on an entire draft in record time and in the most helpful manner.

Jacqui Coughlan, my partner, uncomplainingly lost much of me to this project for 18 months. She absorbed extra housework and endured too many quiet weekday evenings while I wrote about cooperative conflict and loneliness. She also read and offered wonderful insights into every chapter, kept me focused on who I am and why I write, and tirelessly talked the book up to anyone who showed an interest. I could have written a book without her, but it wouldn't be any good.

Notes

WHEN I REFER TO A book by name and the authors in the text, then it appears only in the bibliography and not in these chapter notes.

Prologue

Daniel Dennett (1995) called evolution "the most important idea anybody ever had" in *Darwin's Dangerous Idea*. Nesse and Stearns (2008) provide an excellent introduction to Darwinian medicine. Gould and Lewontin's (1979) attack on sociobiology, adaptationism, just-so stories and Panglossism is in their 1979 paper "The Spandrels of San Marco and the Panglossian Paradigm." The point about just-so stories being hypotheses is from David Queller's (1995) equally erudite and entertaining rebuttal "The Spaniels of St. Marx and the Panglossian Paradox." I first heard the Robert MacArthur quote in a conversation with the eminent ecologist James H. Brown, to whom MacArthur said it. I see Brown (1999) has since published the quote. Michael Shermer's (2007) *The Mind of the Market* is an excellent overview of the common ground between evolution and economics.

1 The weight of our ancestry

The number of obese and overweight people is an interpolation from World Health Organization projections that 2.3 billion adults would be overweight and 700 million obese by 2015 (WHO, 2006). The number of people who are undernourished is from a report by the Food and Agriculture Organization (FAO, 2008). The Jean Ziegler quote regarding deaths due to undernourishment comes from his report to the UN on the right to food (Ziegler, 2001).

Corbett and colleagues (2009) proposed the idea that strong selection on female fertility may be linked, especially via polycystic ovary syndrome and infertility, in ethnic groups experiencing widespread obesity for the first time.

The plant composition of ape diets comes from Milton (2003). Eaton (2006) provides a wonderful paper on the evolution of human diet, and is the source for the benefits of cooking and the idea that hunting co-evolved with the evolution of big brains and bodies. My information on the macronutrient compositions of hunter-gatherer and modern diets come from this paper and two others (Cordain et al., 2000, Eaton et al., 1996).

Tom Standage's book *An Edible History of Humanity* (2009) provides a superb overview of the history of food and is the source for the quote about "the worst mistake in . . . history." I relied on this book and Mark Kurlansky's wonderful *Cod: A Biography of the Fish that Changed the World* (1997) in writing about the industrialization of sugar production and trade. The information about French fries, potato farming and fast food in this chapter and the next comes from Erik Schlosser's *Fast Food Nation* (2001).

The experiments in which caterpillars evolved resistance to obesity on carbohydrate-rich diets are documented by Warbrick-Smith, Behmer, Lee, Raubenheimer and Simpson (2006) and sources therein.

2 Obesity is not for everyone

The links between obesity and poverty, gender, race and education are published in a variety of sources (Drewnowski and Specter, 2004, Paeratakul et al., 2002, McLaren, 2007). An English translation of Kikunae Ikeda's work isolating the chemical basis of umami is now available (Ikeda, 2002).

Data in table 2.1 are standardized obesity data from the World Health Organization's Global Database on Body Mass Index, which includes the results of a large number of surveys and studies. I only show data from surveys post-1998, and for countries where there were multiple surveys post-1998 we used the most recent. Data represent percentage of adult (older than 15 years) men and women with body mass index greater than or equal to 30.0, the standard WHO definition of adult obesity rate.

The Peruvian spider monkey study by the Feltons and their collaborators is in the journal *Behavioral Ecology* (Felton et al., 2009). The experimental study on protein and carbohydrate-rich diets in the Swiss Alps is described by Simpson et al. (2003), and the best description of the protein leverage hypothesis is Simpson and Raubenheimer (2005). The calculation that we eat an extra 53 kilocalorie for every kilocalorie of protein we fail to eat comes from an analysis by Cheng, Simpson and Raubenheimer (2008).

I relied on a number of papers about the links between nutrient density, food pricing and obesity (Drewnowski and Darmon, 2004a, 2004b, Monsivais and Drewnowski, 2009), and an analysis by Christian and Rasad (2009) for trends in food prices. My work with Simpson and Raubenheimer on the price of protein relative to carbohydrates is in *Obesity Reviews* (Brooks et al., 2010).

The figures regarding medical spending on obesity come from Finkelstein et al. (2009). Rangan and co-authors (2007, 2008) discuss the importance of carbohydrate-rich treats in Australians at risk of obesity, and Vartanian et al. (2007) give a US perspective on soft drinks. The data on elasticity of soft-drink demand is from a beverage industry newsletter, cited by Brownell and Frieden (2009), who discuss the moves to tax soft drinks.

McDonald's nutrition information: <nutrition.mcdonalds.com>, accessed 21 July 2011. Prices are from Kingsford, NSW, store on 4 March 2011. Much of the evidence regarding economic and political causes of obesity on Kosrae and elsewhere in the Pacific comes from a paper by Cassels (2006).

3 Weapons of massive consumption

My comments about vegetation damage due to elephants in Kruger National Park are based on an online comment by former ranger Ron Thomson (1988). Owen-Smith et al. (2006) summarize the scientific case for and against elephant culling.

World population estimates come from the US Census Bureau's World Population Clock. The figures on greenhouse gas production by the United States and China come from the World Resources Institute's Climate Analysis Indicators Tool <http:// cait.wri.org>.

The E. O. Wilson quote is from *The Diversity of Life* (1992), pp. 328–29. Garrett Hardin's original "The Tragedy of the Commons" paper was published in *Science* (1968).

4 Dwindling fertility

The George H. W. Bush quote comes from the foreword he wrote to Piotrow's (1973) book.

Evidence that human population dwindled around 70,000 years ago to as few as 2000 comes from mitochondrial DNA evidence (Behar et al., 2008). Estimates of ancient population size come from Davis (1986). The assertion that one-fifth of people live on less than $1 per day is from a report to the World Resources Institute (WRI, 2005).

The idea that in most agricultural societies only relatively wealthy families consistently reproduced and that poverty was an evolutionary dead end comes from Harpending and Cochran (2009). The links between agriculture, the growth of towns and cities, and diseases are explained clearly by Diamond (1998).

Contemporary birth-rate data come from the CIA World Factbook <https://www.cia.gov/library/publications/the-world-factbook>. Historic American birth rates are neatly summarized by Haines (2008).

Malcolm Potts' (2009) analysis of fertility in the Darwin family is in the epilogue to an excellent special volume on population growth. The *Daily Mail* article tracing Darwin's descendants is available online (Dunk and Dennison, 2009).

The classic analysis of !Kung breastfeeding as a contraceptive is found in Koner and Worthmann (1980).

Ruth Mace (2000) reviews the evolutionary basis of human life histories, including the links between wealth, competition and fertility declines. I first started thinking about the links between sexual conflict and demography when I heard Patty Gowaty speak at a conference in Canberra in 1996. Despite regular pestering I have not seen her publish these ideas, which others have touched on but not put forward with the same clarity I recall from her talk.

Sinding (2009) discusses cause and effect in the relationships between wealth and fertility and provides the numbers on unwanted pregnancies and abortions in the developing world. Information on the unmet demand for family planning and abortion in the developing world is in Potts (2009). The example of how education has altered fertility in Iran is described by Lutz (2009).

Michelle Goldberg's (2010) *Means of Reproduction* contains many important insights into the links between the status of women and fertility, as well as analysis of US and Vatican policy on abortion and family planning. I encountered the book after I had written the first draft of this chapter, and was struck by the parallels between her message and what I was trying to achieve here. I have chased down and used several of her quoted sources. The quote at the conclusion of this chapter is from p. 234.

5 Shakespearean love

The mating behavior of male southern elephant seals is described by McCann (1981), although rather more is understood about the behavior of northern elephant seals (e.g. Leboeuf, 1974).

Simon Griffith (2002) reviewed the evidence of extra-pair paternity in birds. The information on superb fairy wrens' fidelity comes from a study by Mike Double and Andrew Cockburn (2000).

For an interesting review of sexual conflict in humans, see Mulder and Rauch (2009).

Alan Dixson (2009) provides a thorough review of estrus in primates. Many of the ideas on the evolution of concealed ovulation and estrus, and men's capacity to detect women's fertility are summarized by Gangestad and Thornhill (2008). The original source for the lap dance study is a paper by Miller, Tybur and Jordan (2007).

Much of the information on the hormonal basis of love is summarized in Zeki's (2007) excellent review.

6 Wrapped around your finger

The Geoff Ogilvy quote comes from an AAP article that appeared in the *Sydney Morning Herald* on 6 January 2010 (Both, 2010).

The classic source on traumatic insemination in bedbugs is Stutt and Siva-Jothy's (2001) original description. Isabella Rosselini enacts, with eccentric brilliance, this trauma at <www.youtube.com/watch?v=MakIB_IJnu0>.

All the Adrienne Germain quotes come from the transcript of an interview by Rebecca Sharples (2003). Amartya Sen's (1986) ideas on cooperative conflicts within families are in his chapter in Irene Tinker's book.

Gurven et al. (2009) model how specialized roles arise as individuals become efficient at particular tasks. Rebecca Bleige Bird's (2007) study on fishing and division of labor among the Meriam is published in *American Anthropologist*. Kristen Hawkes (1990) proposed and popularized the idea that men hunt for show. Hawkes and Bird (2002) review these ideas.

Accounts of life among the Mehinaku are based on Gregor's (1985) book. Information on Aché life and marriage comes from Hill and

Hurtado (1996). My source on partible paternity, especially in the Bari, is Beckerman and Valentine (2002). Gilding (2005) provides a thoughtful re-examination of extra-pair paternity rates in modern humans and the idea that rampant extra-pair paternity is an urban myth.

The best source on the sex contract is Helen Fisher's (1983) book of the same name, or her later book, *The Anatomy of Love* (1992).

Information on the work that men and women do in Australian households is from Baxter et al. (2005). Dempsey's (1997) book gives a sociological overview of inequalities within marriage and whether there is a trend to more equality.

Many of the topics I cover in this chapter are dealt with in Hrdy's (2009) *Mothers and Others*, including the relationships between matrilocality, patrilocality and sexual conflict in primates and people. My example of Iroquois matrilocality and farming comes from Hart (2001), and the impressions among European settlers is from the textbook by Haviland et al. (2007).

7 Love is a battlefield

Two of the best popular books on human mating systems are David Barash and Judith Eve Lipton's (2001) *The Myth of Monogamy*, and Ryan and Jethá's (2010) *Sex at Dawn*. *Anatomy of Love* by Helen Fisher (1992) is also a wonderful read, though it is getting a little dated. The Fisher quote about monogamy is on p. 72.

Evidence that polygyny is more common in foraging societies in the tropics and where food is more abundant, and that polygyny is correlated with violence and war, comes from a review article by Frank Marlowe (2003).

The discussion of sexual coercion and polygyny in Arnhem Land is based on Chisholm and Burbank (1991). The effect of agriculture on wealth inequality comes from Shenk et al. (2009).

I found the Genghis Khan quote in Harpending and Cochran's (2009) book, and that is also my source for the links between wealth and fitness throughout history. The primary source regarding Khan's Y chromosome is Zerjal et al. (2003).

Martin Daly and Margo Wilson's (1988) book *Homicide* gives a good summary of much of their work. The quote about "rare, fatal consequences" is from p. 146. Their work on income inequality in the United States and Canada is published with Shawn Vasdev (Daly et al., 2001).

George Bernard Shaw's ideas on polygyny come from his *Maxims for Revolutionists*, published as an appendix to his play *Man and Superman* (1903). The model showing that women's preferences to maximize resources they can obtain can, alone, explain polygyny are by Kanazawa and Still (1999). Bobbi Low's (1990) original analysis showing that polygyny is commonest when pathogen stress is high is found in her paper in *American Zoologist*. More recently, her review with Martin and Carol Ember shows that both pathogen stress and warfare are associated with polygyny (Ember et al., 2007).

The evidence that children of women in polygynous marriages fare worse than offspring of monogamous unions is published by Jankowiak et al. (2005). The evidence that African women in polygynous marriages fare worse than monogamously married women is presented by Bove and Valeggi (2009).

Richard Alexander's (1979) ideas on polygyny and democracy come from a series of lectures published as the book *Darwinism and Human Affairs*. The Catholic position on marriage comes from an op-ed by Cardinal George Pell (2010) in *The Australian*.

The quotes from Kate Zuma's suicide note as well as quotes by Jacob Zuma are in many news articles. I took them from an article in South Africa's *Sunday Times* (Molele, 2007). The study describing how Mormon polygynists tend to favor a particular wife is by Jankowiak et al. (2005). The Schopenhauer quote is from the essay "On Women" from his *Studies in Pessimism* (1851).

8 Where have all the young girls gone?

I refer to first- and second-wave feminism here and in chapter 10. Feminism's first wave, roughly from 1848 to 1920, resulted in women in many countries gaining the right to vote, to receive formal education, to divorce and to hold property in marriage. The second wave, which reached its peak in the 1970s, opened up workplaces, the professions, government and public institutions as well as challenging the roles of women in the home and family. It also drew attention to sexual double standards and the prevalence of harassment and violence against women.

Sen's (1990) "100 Million Women are Missing" essay appeared in the *New York Review of Books*. Chinese sex ratios are from the 2005 census (Zhu et al., 2009) and Indian numbers come from the 2001 census (Gupta et al., 2002).

Trivers and Willard's (1973) original *Science* paper remains a clear introduction to the Trivers–Willard hypothesis. The example of sex-ratio manipulation by female wasps comes from work by Eric Charnov and his colleagues (1981). The red deer sex-ratio example is from work by Clutton-Brock et al. (1984). The evidence for a Trivers–Willard effect in mammals and the suggestion that glucose might be involved are presented in a meta-analysis by Cameron (2004). Her study of *Forbes* billionaires is in Cameron and Dalerum (2009). The study of sex ratios in polygynous Rwanda is reported by Pollet et al. (2009).

The Leezen example is from Voland (1984) and the Krummhorn example from Voland and Dunbar (1995).

The Michelle Goldberg (2010) quote about dowry is from p. 178 of *The Means of Reproduction* and the translation of *khanya bhronn hatya* comes from p. 184. Information on patrilocality and bride-swapping among Indian villages is from Hudson and den Boer (2004). Many of the insights into Indian and Chinese family life and South Korean son preference come from Gupta et al. (2002).

Insights regarding excess women among cohorts of American baby-boomers come from Pedersen (1991) and from Guttentag and Secord

(1983). Kruger and Schlemmer (2009) analyzed sex-ratio effects on marriage in contemporary American cities and the 1910 study of American states is by Pollet and Nettle (2008).

The article mentioning Baljeet Singh and Sona Katum was published in *The Economist*, 6 March 2010. Articles in this issue first alerted me to the missing-women problem and to some of my main sources.

9 Blame it on the Stones

A. E. Hotchner (1990) alleges in *Blown Away: The Rolling Stones and the Death of the Sixties* that Jones was killed at a party at his mansion, probably by builders who had been working on his house and who were attending the party as Jones' guests. Several journalists have added further evidence in support of Hotchner's allegation, and in 2009 the Sussex police reopened the investigation into Jones' death.

The Germaine Greer (1970) quote is from *The Female Eunuch*. All the Keith Richards quotes come from his biography, *Life* (Richards and Fox, 2010). The Geoffrey Miller (2000) quote is from a book chapter on music, which influenced a number of the ideas I present in this chapter.

The motion-capture and animation study of dancing men is by Neave et al. (2010). Janssen's (2007) work on the reminiscence bump in music is published in the journal *Memory*. The original source for the "music hath charms to soothe the savage breast" is William Congreve's *The Mourning Bride* (1697).

Rentfrow and Gosling's study on how young adults use music in getting to know one another is in their 2006 paper, and their work on musical tastes and the Big Five personality traits is in their 2003 paper. The idea that musical omnivores are more socially adept and likely to attain higher status comes from Tanner et al. (2008). The role of music in the lives of the elderly comes from an Australian study (Hays and Minichiello, 2005). The Timothy Garton Ash (2001) quote is from the essay "Marta and Helena" in *History of the Present*.

10 About a boy

The Matt Cameron quote is from his introduction to the chapter "Grunge" in *Rolling Stone*'s book *The '90s: The Inside Stories from the Decade that Rocked.*

Millar (2008) presents evidence that boys and girls don't differ in musical talent or likelihood of beginning with an instrument. All analyses of *Billboard* number ones come from my own research in the public domain. The data on women singers before and after 1954 come from Wilkinson (1976).

Michael Kasumovic's (2011) experiments on field crickets are in press at *Journal of Evolutionary Biology.*

The Pinker (2002) quote is from *The Blank Slate*, and I have drawn on that exceptional book in my characterization of social constructionism and gender feminism.

The Shirley Manson (2004) quote appeared in *Rolling Stone*. Elijah Wald (2009) quotes come from *How the Beatles Destroyed Rock 'n' Roll*. The subheading "It crawled from the South" comes from the title of Marcus Gray's excellent book on R.E.M. Michael Ventura's (1985) entertaining essay "Hear that Long Snake Moan" is from a volume that is out of print, but it can be downloaded from <www.michaelventura.org/writings/EB2.pdf>.

Assertions regarding Elvis' FBI files come from Fensch (2001). The Camille Paglia (1992) quote is from p. 59 of *Sex, Art, and American Culture.* Gary Smith's (1996) *Sports Illustrated* article is the source of the Earl Woods quote. The quote about Muhammad Ali's press conference comes from Hauser's (2004) biography of Ali. The Daly and Wilson (1988) quote is from *Homicide.* Gary Herman's (1982) groupie quote comes from p. 115 and the Deborah Harry quote from p. 126 of *Rock 'n' Roll Babylon.*

The analysis of fitness among Lamalera whale-hunters comes from Smith (2004). Packer et al. (1991) discuss coalition formation among related lions. Conner et al. (1992) discuss dolphin alliances, and Krützen et al. (2003) show alliance members tend to be related.

Anthony Bozza's (2009) quote is from *Why AC/DC Matters*. The Vivien Johnson (1992) quote is from a chapter she wrote for Hayward's book about popular music in Australia. Evidence that women drop out of the audience as the rock gets harder comes from Millar (2008).

11 Immortality

The Hunter S. Thompson line is a misquote with a life of its own, a mutation in the DNA of the interwebs. As David Emery explains <http://urbanlegends.about.com/od/dubiousquotes/a/hunter_thompson.htm>, it is one of many distortions of something Thompson said on p. 43 of *Generation of Swine* (1988) regarding the television business.

Australian causes of death (AIHW, 2006) and cancer statistics (AIHW, 2010) come from the Australian Institute of Health and Welfare. Daniel Martinez is quoted in *Pomona College Magazine* (March 2009). His paper on the lack of aging in *Hydra* was published in *Experimental Gerontology* (Martinez, 1998).

George Williams' (1957) classic paper on the role of pleiotropy in aging remains a clear and incisive piece of thinking on the topic. Callaway's (2010) piece in *New Scientist* is a useful introduction to the trade-off between Alzheimer's and early intelligence, and it is the source of my claims about James Watson, Stephen Pinker and Craig Venter. I also relied on the papers by Hubacek et al. (2001) and Marchant et al. (2010). Westerdorp and Kirkwood (1998) published the analysis of British aristocrats. Maklakov's (2008) paper is in *Evolution and Human Behavior*. Nettle's (2010) study on modern English families is in *Behavioral Ecology*. The comparison of East Germany and West Germany is from Hank (2010).

The best reference for field crickets that live fast and die young is Hunt et al. (2004). Moore and Wilson (2002) is a good place to start for the links between infectious disease and male–male competition intensity. Kruger and Nesse's (2007) study of Eastern European transitions to market economies is in *Evolutionary Psychology.* The graph of the mortality of *Rolling Stone* Immortals comes from my own analysis using publicly available biographic information, especially Wikipedia. The analysis of the musicians who made the Top 1000 albums comes from Bellis et al. (2007).

The Jim Morrison quote is on the Doors album *American Prayer*, straight after a live version of "Road House Blues." The quote about Robert Johnson is from *Hammer of the Gods*, Stephen Davis' (1985) book about Led Zeppelin. The Ozzy Osbourne (2010) quote comes from a *Sunday Telegraph* article attributed to him.

"It's better to burn out than to fade away" and "once you're gone, you can never come back" are both from Neil Young's "My My, Hey Hey (Out of the Blue)" from the album *Rust Never Sleeps.* The song is published by Silver Fiddle Music. Words and music are by Neil Young and Jeff Blackburn.

Bibliography

Australian Institute of Health and Welfare (2006) *Mortality over the Twentieth Century in Australia: Trends and Patterns in Major Causes of Death*, Canberra, Australian Institute of Health and Welfare.

— (2010) Australian Cancer Database, Canberra, Australian Institute of Health and Welfare.

Alexander, R. D. (1979) *Darwinism and Human Affairs*, Seattle, WA, University of Washington Press.

Barash, D. P. & Lipton, J. E. (2001) *The Myth of Monogamy: Fidelity and Infidelity in Animals and People*, New York, W.H. Freeman and Company.

Basu, A. M. (1992) *Culture, the Status of Women, and Demographic Behaviour*, Oxford, Oxford University Press.

Baxter, J., Hewitt, B. & Western, M. (2005) Post-familial families and the domestic division of labour, *Journal of Comparative Family Studies*, 36, 583–604.

Beckerman, S. & Valentine, P. (2002) *The Theory and Practice of Partible Paternity in South America*, Gainesville, FL, University Press of Florida.

Behar, D. M., Villems, R., Soodyall, H., Blue-Smith, J., Pereira, L., Metspalu, E., Scozzari, R., Makkan, H., Tzur, S., Comas, D., Bertranpetit, J., Quintana-Murci, L., Tyler-Smith, C., Wells, R. S. & Rosset, S. (2008) The dawn of human matrilineal diversity, *American Journal of Human Genetics*, 82, 1130–40.

Bellis, M. A., Hennell, T., Lushey, C., Hughes, K., Tocque, K. & Ashton, J. R. (2007) Elvis to Eminem: Quantifying the price of fame through early mortality of European and North American rock and pop stars, *Journal of Epidemiology and Community Health*, 61.

Bird, R. B. (2007) Fishing and the sexual division of labor among the Meriam, *American Anthropologist*, 109, 442–51.

Both, A. (2010) Fans will want Tiger back: Ogilvy, *Sydney Morning Herald*, Sydney, Fairfax Media.

Bove, R. & Valeggia, C. (2009) Polygyny and women's health in sub-Saharan Africa, *Social Science and Medicine*, 68, 21–29.

Bozza, A. (2009) *Why AC/DC Matters*, New York, William Morrow.

Brooks, R. C., Simpson, S. J. & Raubenheimer, D. (2010) The price of protein: Combining evolutionary and economic analysis to understand excessive energy consumption, *Obesity Reviews*, 11, 887–94.

Brown, J. H. (1999) The legacy of Robert MacArthur: From geographical ecology to macroecology, *Journal of Mammalogy*, 80, 333–44.

Brownell, K. D. & Frieden, T. R. (2009) Ounces of prevention—the public policy case for taxes on sugared beverages, *New England Journal of Medicine*, 360, 1805–808.

Callaway, E. (2010) Alzheimer's gene makes you smart, *New Scientist*, 13 February, 12–13.

Cameron, E. Z. (2004) Facultative adjustment of mammalian sex ratios in support of the Trivers–Willard hypothesis: Evidence for a mechanism, *Proceedings of the Royal Society, Series B, Biological Sciences*, 271, 1723–28.

Cameron, E. Z. & Dalerum, F. (2009) A Trivers–Willard effect in contemporary humans: Male-biased sex ratios among billionaires, *PLoS ONE*, 4.

Cassels, S. (2006) Overweight in the Pacific: Links between foreign dependence, global food trade, and obesity in the Federated States of Micronesia, *Global Health*, 2, 10.

Charnov, E. L., Los-den Hartogh, R. L., Jones, W. T. & van den Assem, J. (1981) Sex ratio evolution in a variable environment, *Nature*, 289, 27.

Cheng, K., Simpson, S. J. & Raubenheimer, D. (2008) A geometry of regulatory scaling, *The American Naturalist*, 172, 681–93.

Chisholm, J. S. & Burbank, V. K. (1991) Monogamy and polygyny in southeast Arnhem Land: Male coecion and female choice, *Ethology and Sociobiology*, 12, 291–313.

Christian, T. & Rashad, I. (2009) Trends in US food prices, 1950–2007, *Economics and Human Biology*, 7, 113–20.

Clutton-Brock, T., Albon, S. D. & Guinness, F. E. (1984) Maternal dominance, breeding success and birth sex ratios in red deer, *Nature*, 308, 358–60.

Connor, R. C. (1992) Two levels of alliance formation among male bottlenose dolphins (*Tursiops* sp.), *Proceedings of the National Academy of Sciences of the United States of America*, 89, 987–90.

Corbett, S. J., McMichael, A. J. & Prentice, A. M. (2009) Type 2 diabetes, cardiovascular disease, and the evolutionary paradox of the polycystic ovary syndrome: A fertility first hypothesis, *American Journal of Human Biology*, 21, 587–98.

Cordain, L., Miller, J. B., Eaton, S. B. & Mann, N. (2000) Macronutrient estimations in hunter-gatherer diets, *American Journal of Clinical Nutrition*, 72, 1589–90.

Daly, M. & Wilson, M. (1988) *Homicide*, New Brunswick, NJ, Transaction Publishers.

Daly, M., Wilson, M. & Vasdev, N. (2001) Income inequality and homicide rates in Canada and the United States, *Canadian Journal of Criminology*, 43, 219–36.

Darwin, C. (1859) *On the Origin of Species by Means of Natural Selection or the Preservation of Favoured Races in the Struggle for Life*, London, Murray.

—— (1871) *The Descent of Man, and Selection in Relation to Sex*, London, Murray.

Davis, K. (1986) The history of birth and death, *Bulletin of the Atomic Scientists*, 42, 20–23.

Davis, S. (1985) *Hammer of the Gods: The Led Zeppelin Saga*, New York, William Morrow & Co.

Dempsey, K. (1997) *Inequalities in Marriage: Australia and Beyond*, Melbourne, Oxford University Press.

Dennett, D. C. (1995) *Darwin's Dangerous Idea: Evolution and the Meanings of Life*, New York, Simon and Schuster.

Diamond, J. (1998) *Guns, Germs and Steel: A Short History of Everybody for the Last 13,000 Years*, London, Vintage.

—— (2005) *Collapse: How Societies Choose to Fail or Succeed*, New York, Viking.

Dixson, A. F. (2009) *Sexual Selection and the Evolution of Animal Mating Systems*, New York, Oxford University Press.

Double, M. & Cockburn, A. (2000) Pre-dawn infidelity: Females control extra-pair mating in superb fairy-wrens, *Proceedings of the Royal Society of London, Series B, Biological Sciences*, 267, 465–70.

Drewnowski, A. & Darmon, N. (2004a) The economics of obesity: Dietary energy density and energy cost, *Symposium on Science-based Solutions to Obesity*, Anaheim, CA.

—— (2004b) Food choices and diet costs: An economic analysis, *Symposium on Modifying the Food Environment*, Washington, DC.

Drewnowski, A. & Specter, S. E. (2004) Poverty and obesity: The role of energy density and energy costs, *American Journal of Clinical Nutrition*, 79, 6–16.

Dunk, M. & Dennison, M. (2009) The descent of man: We trace those who claim Charles Darwin as an ancestor, *Daily Mail*, London, Associated Newspapers Ltd.

Eaton, S. B. (2006) The ancestral human diet: What was it and should it be a paradigm for contemporary nutrition? *Proceedings of the Nutrition Society*, 65, 1–6.

Eaton, S. B., Konner, M. J. & Shostak, M. (1996) An evolutionary perspective enhances understanding of human nutritional requirements, *Journal of Nutrition*, 126, 1732–40.

Ember, M., Ember, C. R. & Low, B. S. (2007) Comparing explanations of polygyny, *Cross-Cultural Research*, 41, 428–40.

Food and Agriculture Organization (2008) *The State of Food Insecurity in the World 2008*, Rome, Food and Agriculture Organization of the United Nations.

Farber, B. A. (2007) *Rock 'n' Roll Wisdom: What Psychologically Astute Lyrics Teach about Life and Love*, Westport, CT, Praeger Publishers.

Felton, A. M., Felton, A., Raubenheimer, D., Simpson, S. J., Foley, W. J., Wood, J. T., Wallis, I. R. & Lindenmayer, D. B. (2009) Protein content of diets dictates the daily energy intake of a free-ranging primate, *Behavioral Ecology*, 20, 685–90.

Fensch, T. (2001) *The FBI Files on Elvis Presley*, The Woodlands, TX, New Century Books.

Finkelstein, E. A., Trogdon, J. G., Cohen, J. W. & Dietz, W. (2009) Annual medical spending attributable to obesity: Payer- and service-specific estimates, *Health Affairs*, 28, w822–31.

Fisher, H. (1983) *The Sex Contract—The Evolution of Human Behavior*, New York, William Morrow & Co.

—— (1992) *Anatomy of Love*, New York, Fawcett Columbine.

Gangestad, S. W. & Thornhill, R. (2008) Human oestrus, *Proceedings of the Royal Society of London, Series B, Biological Sciences*, 275, 991–1000.

Garton Ash, T. (2001) *History of the Present: Essays, Sketches, and Dispatches from Europe in the 1990s*, New York, Vintage.

Gilding, M. (2005) Rampant misattributed paternity: The creation of an urban myth, *People and Place*, 13, 1–11.

Goldberg, M. (2010) *The Means of Reproduction: Sex, Power and the Future of the World*, London, Penguin Books.

Gould, S. J. & Lewontin, R. C. (1979) The spandrels of San Marco and the Panglossian paradigm: a critique of the adaptionist programme, *Proceedings of the Royal Society of London, Series B, Biological Sciences*, 205, 581–98.

Greer, G. (1970) *The Female Eunuch*, London, MacGibbon and Kee.

Gregor, T. (1985) *Anxious Pleasures: The Sexual Lives of an Amazonian People*, Chicago, University of Chicago Press.

Griffith, S. C. (2002) Extra pair paternity in birds: A review of interspecific variation and adaptive function, *Molecular Ecology*, 11, 2195–212.

Gupta, M. D., Zhenghua, J., Li Bohua, X. Z. & Woojin Chung, B. H.-O. (2002) Why is son preference so persistent in East and South Asia? A cross-country study of China, India and the Republic of Korea, *World Bank Policy Research Working Paper No. 2942*, available at http://ssrn.com/abstract=636304, Washington, The World Bank.

Gurven, M., Winking, J., Kaplan, H., Von Rueden, C. & McAllister, L. (2009) A bioeconomic approach to marriage and the sexual division of labor, *Human Nature*, 20, 151–83.

Guttentag, M. & Secord, P. (1983) *Too Many Women? The Sex Ratio Question*, Beverly Hills, Sage.

Haines, M. (2008) Fertility and mortality in the United States, in Whaples, R. (ed.), *EH.Net Encyclopedia*.

Hank, K. (2010) Childbearing history, later-life health, and mortality in Germany, *Population Studies—A Journal of Demography*, 64, 275–91.

Hardin, G. (1968) The Tragedy of the Commons, *Science*, 162, 1243–48.

Harpending, H. & Cochran, G. (2009) *The 10,000 Year Explosion: How Civilization Accelerated Human Evolution*, New York, Basic Books.

Hart, J. P. (2001) Maize, matrilocality, migration and Northern Iroquoian evolution, *Journal of Archaeological Method and Theory*, 8, 151–82.

Hauser, T. (2004) *Muhammad Ali: His Life and Times*, London, Robson Books.

Haviland, W. A., Prins, H. E. L., Walrath, D. & McBride, B. (2007) *Cultural Anthropology: The Human Challenge*, Belmont, CA, Thomson Learning.

Hawkes, K. (1990) Why do men hunt? Benefits for risky choices, in Cashdan, E. (ed.), *Risk and Uncertainty in Tribal and Peasant Economies*, Boulder, CO, Westview Press.

Hawkes, K. & Bird, R. B. (2002) Showing off, handicap signaling, and the evolution of men's work, *Evolutionary Anthropology*, 11, 58–67.

Hays, T. & Minichiello, V. (2005) The contribution of music to quality of life in older people: An Australian qualitative study, *Ageing and Society*, 25, 261–78.

Herman, G. (1982) *Rock 'n' Roll Babylon*, London, Plexus Publishing.

Hill, K. & Hurtado, A. M. (1996) *The Ecology and Demography of a Foraging People*, New York, Aldine de Gruyer.

Holland, B. & Rice, W. R. (1999) Experimental removal of sexual selection reverses intersexual antagonistic coevolution and removes a reproductive load, *Proceedings of the National Academy of Sciences of the United States of America*, 96, 5083–88.

Hotchner, A. E. (1990) *Blown Away: The Rolling Stones and the Death of the Sixties*, New York, Simon & Schuster.

Hrdy, S. B. (2009) *Mothers and Others: The Evolutionary Origins of Mutual Understanding*, Cambridge, MA, Harvard University Press.

Hubacek, J., Pitha, J., Skodová, Z., Adámková, V., Lánská, V. & Poledne, R. (2001) A possible role of apolipoprotein E polymorphism in predisposition to higher education, *Neuropsychobiology*, 200–203.

Hudson, V. M. & den Boer, A. M. (2004) *Bare Branches: The Security Implications of Asia's Surplus Male Population*, Cambridge, MA, MIT Press.

Hunt, J., Brooks, R., Smith, M. J., Jennions, M. D., Bentsen, C. L. & Bussière, L. F. (2004) High quality male field crickets invest heavily in sexual display but die young, *Nature*, 432, 1024–27.

Ikeda, K. (2002) New seasonings, *Chemical Senses*, 27, 847–49.

Jankowiak, W., Sudakov, M. & Wilreker, B. C. (2005) Co-wife conflict and co-operation, *Ethnology*, 44, 81–98.

Janssen, S. M. J., Chessa, A. G. & Murre, J. M. J. (2007) Temporal distribution of favourite books, movies, and records: Differential encoding and re-sampling, *Memory*, 15, 755–67.

Johnson, V. (1992) Be my woman rock 'n' roll, in Hayward, P. (ed.), *From Pop to Punk to Postmodernism: Popular Music and Australian Culture from the 1960s to the 1990s*, North Sydney, Allen & Unwin.

Kanazawa, S. & Still, M. C. (1999) Why monogamy? *Social Forces*, 78, 25–50.

Kasumovic, M. M., Hall, M. D., Try, H. & Brooks, R. C. (2011) The importance of listening, *Journal of Evolutionary Biology*, 24, 1325–1334.

Konner, M. & Worthman, C. (1980) Nursing frequency, gonadal-function, and birth spacing among !Kung hunter-gatherers, *Science*, 207, 788–91.

Krützen, M., Sherwin, W. B., Connor, R. C., Barre, L. M., Van de Casteele, T., Mann, J. & Brooks, R. (2003) Contrasting evolutionary strategies within a population of bottlenose dolphins (*Tursiops* sp.), *Proceedings of the Royal Society of London, Series B, Biological Sciences*, 270, 497–502.

Kruger, D. J. & Nesse, R. M. (2007) Economic transition, male competition and sex differences in mortality, *Evolutionary Psychology*, 5, 411–27.

Kruger, D. J. & Schlemmer, E. (2009) Male scarcity is differentially related to male marital likelihood across the life course, *Evolutionary Psychology*, 7, 280–87.

Kurlansky, M. (1997) *Cod: A Biography of the Fish that Changed the World*, New York, Walker Publishing Inc.

Lawrence, P. R. & Nohria, N. (2002) *Driven: How Human Nature Shapes our Choices*, San Francisco, CA, Jossey-Bass.

Leboeuf, B. J. (1974) Male–male competition and reproductive success in elephant seals, *American Zoologist*, 14, 163–76.

Levitt, S. D. & Dubner, S. J. (2005) *Freakonomics: A Rogue Economist Explores the Hidden Side of Everything*, New York, William Morrow.

Low, B. S. (1990) Marriage systems and pathogen stress in human societies, *American Zoologist*, 30, 325–39.

Lutz, W. (2009) Sola schola et sanitate: Human capital as the root cause and priority for international development? *Philosophical Transactions of the Royal Society of London, Series B, Biological Sciences*, 364, 3031–47.

Ma, J. (2009) The immortal *Hydra*, *Pomona College Magazine*, 45: www.pomona.edu/Magazine/PCMWin09/NKdanielmartinez.shtml.

McCann, T. S. (1981) Aggression and sexual-activity of male southern elephant seals, *Mirounga leonina*, *Journal of Zoology*, 195, 295–310.

Mace, R. (2000) Evolutionary ecology of human life history, *Animal Behaviour*, 59, 1–10.

McLaren, L. (2007) Socioeconomic status and obesity, *Epidemiologic Reviews*, 29, 29–48.

Maklakov, A. A. (2008) Sex difference in life span affected by female birth rate in modern humans, *Evolution and Human Behavior*, 29, 444–49.

Malthus, T. R. (1798) *An Essay on the Principle of Population*, London, J. Johnson.

Manson, S. (2004) The Immortals—The Greatest Artists of All Time: 47 (Patti Smith), *Rolling Stone*, 15 April, 138

Marchant, N. L., King, S. L., Tabet, N. & Rusted, J. M. (2010) Positive effects of cholinergic stimulation favor young APOE 4 carriers, *Neuropsychopharmacology*, 35, 1090–96.

Marlowe, F. W. (2003) The mating system of foragers in the standard cross-cultural sample, *Cross-Cultural Research*, 37, 282–306.

Martinez, D. E. (1998) Mortality patterns suggest lack of senescence in Hydra, *Experimental Gerontology*, 33, 217–25.

Millar, B. (2008) Selective hearing: Gender bias in the music preferences of young adults, *Psychology of Music*, 36, 429–45.

Miller, G. F. (2000) Evolution of human music through sexual selection, in Wallin, N. L., Merker, B. & Brown, S. (eds.), *The Origins of Music*, Boston, MIT Press.

—— (2001) *The Mating Mind*, New York, Anchor Books.

Miller, G., Tybur, J. M. & Jordan, D. J. (2007) Ovulatory cycle effects on tip earnings by lap dancers: Economic evidence for human estrus? *Evolution and Human Behavior*, 28, 375–81.

Milton, K. (2003) The critical role played by animal source foods in human (*Homo*) evolution, *Journal of Nutrition*, 133, 3886S–3892S.

Molele, C. (2007) South Africa: So who will the polygamous Zuma's First Lady be? London, *Sunday Times*, 23 December.

Monsivais, P. & Drewnowski, A. (2009) Lower-energy-density diets are associated with higher monetary costs per kilocalorie and are consumed by women of higher socioeconomic status, *Journal of the American Dietetic Association*, 109, 814–22.

Moore, S. L. & Wilson, K. (2002) Parasites as a viability cost of sexual selection in natural populations of mammals, *Science*, 297, 2015–18.

Mulder, M. B. & Rauch, K. L. (2009) Sexual conflict in humans: Variations and solutions, *Evolutionary Anthropology*, 18, 201–14.

Neave, N., McCarty, K., Freynik, J., Caplan, N., Hönekopp, J. & Fink, B. (2010) Male dance moves that catch a woman's eye, *Biology Letters*, doi: 10.1098/rsbl.2010.0619.

Nesse, R. M. & Stearns, S. C. (2008) The great opportunity: Evolutionary applications to medicine and public health, *Evolutionary Applications*, 1, 28–48.

Nettle, D. (2010) Dying young and living fast: Variation in life history across English neighborhoods, *Behavioral Ecology*, 21.

Osbourne, O. (2010) Deepest secrets of the Prince of Darkness, Sydney, *Sunday Telegraph*, 31 October.

Owen-Smith, N., Kerley, G. I. H., Page, B., Slotow, R. & Van Aarde, R. J. (2006) A scientific perspective on the management of elephants in the Kruger National Park and elsewhere, *South African Journal of Science*, 102, 389–94.

Packer, C., Gilbert, D. A., Pusey, A. E. & O'Brien, S. J. (1991) A molecular genetic analysis of kinship and cooperation in African lions, *Nature*, 351, 562–65.

Paeratakul, S., Lovejoy, J. C., Ryan, D. H. & Bray, G. A. (2002) The relation of gender, race and socioeconomic status to obesity and obesity comorbidities in a sample of US adults, *International Journal of Obesity*, 26, 1205–10.

Paglia, C. (1992) *Sex, Art, and American Culture*, New York, Vintage Books.

Pedersen, F. A. (1991) Secular trends in human sex ratios: Their influence on individual and family behaviour, *Human Nature*, 2, 271–91.

Pell, G. (2010) Relationships market after 50 years on the Pill, *Australian*, News Limited, 25 September.

Pinker, S. (1997) *How the Mind Works*, New York, W.W. Norton and Company.

—— (2002) *The Blank Slate: The Modern Denial of Human Nature*, New York, Viking.

Piotrow, P. T. (1973) *World Population Crisis: The United States Response*, New York, Praeger Publishers.

Pollan, M. (2008) *In Defense of Food: An Eater's Manifesto*, New York, Penguin Press.

Pollet, T. V., Fawcett, T. W., Buunk, A. P. & Nettle, D. (2009) Sex-ratio biasing towards daughters among lower-ranking co-wives in Rwanda, *Biology Letters*, 5, 765–68.

Pollet, T. V. & Nettle, D. (2008) Driving a hard bargain: Sex ratio and male marriage success in a historical US population, *Biology Letters*, 4, 31–33.

Potts, M. (2009) Where next? *Philosophical Transactions of the Royal Society, Series B, Biological Sciences*, 364.

Queller, D. C. (1995) The spaniels of St. Marx and the Panglossian paradox: A critique of a rhetorical programme, *The Quarterly Review of Biology*, 70 485–89.

Rangan, A., Hector, D., Randall, D., Gill, T. & Webb, K. (2007) Monitoring consumption of "extra" foods in the Australian diet: Comparing two sets of criteria for classifying foods as "extras," *Nutrition & Dietetics*, 64, 261–67.

Rangan, A. M., Randall, D., Hector, D. J., Gill, T. P. & Webb, K. L. (2008) Consumption of "extra" foods by Australian children: Types, quantities and contribution to energy and nutrient intakes, *European Journal of Clinical Nutrition*, 62, 356–64.

Rentfrow, P. J. & Gosling, S. D. (2003) The do re mi's of everyday life: The structure and personality correlates of music preferences, *Journal of Personality and Social Psychology*, 84, 1236–56.

—— (2006) Message in a ballad—The role of music preferences in interpersonal perception, *Psychological Science*, 17, 236–42.

Richards, K. & Fox, J. (2010) *Life*, London, Little, Brown and Company.

Ryan, C. & Jethá (2010) *Sex at Dawn: The Prehistoric Origins of Modern Sexuality*, Melbourne, Scribe.

Schlosser, E. (2001) *Fast Food Nation: The Dark Side of the American Meal*, Boston, Houghton Mifflin.

Schopenhauer, A. ([1851] 2010) *Studies in Pessimism*, Adelaide, ebooks@Adelaide.

Sen, A. (1986) Gender and cooperative conflicts, in Tinker, I. (ed.), *Persistent Inequalities: Women and World Development*, Oxford, Oxford University Press.

—— (1990) More than 100 million women are missing, *New York Review of Books*, 37.

Sharples, R. (2003) transcript of interview with Adrienne Germain, *Population and Reproductive Health Oral History Project*, Northampton, MA, Sophia Smith Collection, Smith College.

Shaw, G. B. (1903) *Man and Superman*, Westminster, Archibald Constable & Co.

—— (1908) *Getting Married*, public domain; available from Project Gutenberg.

Shenk, M. K., Mulder, M. B., Beise, J., Clark, G., Irons, W., Leonetti, D., Low, B. S., Bowles, S., Hertz, T., Bell, A. & Piraino, P. (2009) Intergenerational wealth transmission among agriculturalists: Foundations of agrarian inequality, *Current Anthropology*, 51, 65–83.

Shermer, M. (2007) *The Mind of the Market: Compassionate Apes, Competitive Humans and Other Tales from Evolutionary Economics*, New York, Times Books.

Simpson, S. J., Batley, R. & Raubenheimer, D. (2003) Geometric analysis of macronutrient intake in humans: The power of protein? *Appetite*, 41, 123–40.

Simpson, S. J. & Raubenheimer, D. (2005) Obesity: The protein leverage hypothesis, *Obesity Reviews*, 6, 133–42.

Sinding, S. W. (2009) Population, poverty and economic development, *Philosophical Transactions of the Royal Society, Series B, Biological Sciences*, 364, 3023–30.

Smith, A. (1776) *An Inquiry into the Nature and Causes of the Wealth of Nations*, London, W. Strahan & T. Cadell.

Smith, E. A. (2004) Why do good hunters have higher reproductive success? *Human Nature*, 15, 342–63.

Smith, G. (1996) The chosen one, *Sports Illustrated*, 23 December, 31.

Standage, T. (2009) *An Edible History of Humanity*, New York, Walker.

Stutt, A. D. & Siva-Jothy, M. T. (2001) Traumatic insemination and sexual conflict in the bed bug *Cimex lectularius*, *Proceedings of the National Academy of Sciences of the United States of America*, 98, 5683–87.

Tanner, J., Asbridge, M. & Wortley, S. (2008) Our favourite melodies: Musical consumption and teenage lifestyles, *British Journal of Sociology*, 59, 117–44.

Thompson, H. S. (1988) *Generation of Swine: Tales of Shame and Degradation in the '80s*, New York, Summit Books.

Thomson, R. (2009) On elephant numbers in Kruger, Siyabona Africa Travel (Pty) Ltd, www.krugerpark.co.za/krugerpark-times-23-elephant-numbers-18006.html, accessed 13 September.

Trivers, R. L. & Willard, D. E. (1973) Natural selection of parental ability to vary the sex ratio of offspring, *Science*, 179, 90–92.

Vartanian, L. R., Schwartz, M. B. & Brownell, K. D. (2007) Effects of soft drink consumption on nutrition and health: A systematic review and meta-analysis, *American Journal of Public Health*, 97, 667–75.

Ventura, M. (1985) Hear that long snake moan, *Shadow Dancing in the USA*, Los Angeles, Tarcher's/St. Martin's Press.

Voland, E. (1984) Human sex-ratio manipulation: Historic data from a German parish, *Journal of Human Evolution*, 13, 99–107.

Voland, E. & Dunbar, R. I. M. (1995) Resource competition and reproduction —The relationship between economic and parental strategies in the Krummhorn population (1720–1874), *Human Nature*, 6, 33–49.

Voltaire (1759) *Candide: or, All for the Best*, Paris, Sirène.

Wald, E. (2009) *How the Beatles Destroyed Rock 'n' Roll: An Alternative History of American Popular Music*, New York, Oxford University Press.

Warbrick-Smith, J., Behmer, S. T., Lee, K. P., Raubenheimer, D. & Simpson, S. J. (2006) Evolving resistance to obesity in an insect, *Proceedings of the National Academy of Sciences of the United States of America*, 103, 14045–49.

Westendorp, R. G. J. & Kirkwood, T. B. L. (1998) Human longevity at the cost of reproductive success, *Nature*, 396, 743–46.

World Health Organization (2006) *World Health Organization Fact Sheet No. 311: Obesity and Overweight*, Geneva, World Health Organization.

Wilkinson, M. (1976) Romantic love: The great equalizer? Sexism in popular music, *The Family Coordinator*, April, 161–66.

Williams, G. C. (1957) Pleiotropy, natural selection, and the evolution of senescence, *Evolution*, 11, 398–411.

Wilson, E. O. (1992) *The Diversity of Life*, New York, W.W. Norton and Company.

World Resources Institute (WRI) in collaboration with United Nations Development Programme, United Nations Environment Programme, and World Bank (2005) *World Resources 2005: The Wealth of the Poor—Managing Ecosystems to Fight Poverty*, Washington, DC, WRI.

Xinran (2010) *Message from an Unknown Chinese Mother: Stories of Loss and Love*, Sydney, Random House Australia.

Zeki, S. (2007) The neurobiology of love, *FEBS Letters*, 581, 2575–79.

Zerjal, T., Xue, Y. L., Bertorelle, G., Wells, R. S., Bao, W. D., Zhu, S. L., Qamar, R., Ayub, Q., Mohyuddin, A., Fu, S. B., Li, P., Yuldasheva, N., Ruzibakiev, R., Xu, J. J., Shu, Q. F., Du, R. F., Yang, H. M., Hurles, M. E., Robinson, E., Gerelsaikhan, T., Dashnyam, B., Mehdi, S. Q. & Tyler-Smith, C. (2003) The genetic legacy of the Mongols, *American Journal of Human Genetics*, 72, 717–21.

Zhu, W. X., Lu, L. & Hesketh, T. (2009) China's excess males, sex selective abortion, and one child policy: Analysis of data from 2005 national intercensus survey, *British Medical Journal*, 338: b1211.

Ziegler, J. (2001) *The right to food*, Report by the Special Rapporteur on the right to food, Mr. Jean Ziegler, submitted in accordance with Commission on Human Rights resolution 2000/10, Geneva, United Nations.

Index